设计未来系列丛书

全球设计马拉松
GLOBAL DESIGN DAY MARATHON
可持续的青年设计力
SUSTAINABLE YOUTH DESIGN FORCE

丁肇辰　著

by Chawchen Ting

中国纺织出版社有限公司

内 容 提 要

本书记录了在2021年和2022年设计马拉松期间产出的优秀作品成果。通过这些作品展示，读者们可以了解设计马拉松的成就和其对设计教育与产业发展的贡献。这些作品展现了设计马拉松参与者的创造力和创新思维，为未来的设计发展提供了有价值的参考。

全书图文并茂，内容翔实丰富，图片精美，实践性高、针对性强，具有较高的学习和研究价值，既适合高等艺术院校专业师生学习，也可供相关从业人员和研究者参考使用。

图书在版编目（CIP）数据

全球设计马拉松·可持续的青年设计力：汉文、英
肇辰著. --北京：中国纺织出版社有限公司，
4. --（设计未来系列丛书）. -- ISBN 978-7
2390-1

S941.28

国家版本馆CIP数据核字第2024QU0954号

春奕　施 琦　　责任校对：高　涵
丽

限公司出版发行
区百子湾东里 A407 号楼　邮政编码：100124
04422　传真：010—87155801

见店

2119887771

刷　各地新华书店经销

刷

长：9.25

元

页，由本社图书营销中心调换

北京设计周期间最时尚的设计大赛之一

ONE OF THE MOST FASHIONABLE DESIGN COMPETITION DURING BEIJING DESIGN WEEK

哈佛大学教育研究员克里斯·达德（Chris Dede）曾于2009年提出"面向21世纪的技能"，强调了"技能"具备的广度由技巧、能力及学习倾向构成，由此也可看出，这些"技能"是学生在特定环境下通过各种途径获得和掌握的知识与技术。而"技能"又可以分为感知型、操作型及应用型三种类型，具备这些技能的人才具有较高的智力水平和工作适应性，他们能有效地完成自己承担的任务并对他人做出积极回应。这些年来，我们能从高校招生方式的改变看出学校在培养设计师所应具备的多样性专业技能上的需求不断增加，在学生们进入学校后的设计教学规划上，除了须不断增强学生们的"T型"横向技能外，也非常重视学生们的合作能力，旨在将过去强调"个人的"学习环境过渡到适应"团队的"学习环境。

但是，尽管当前的设计教学模式试图在多专业、多样化的培养设想下建立，我们的设计课堂仍存在以下几个问题：课堂成果形式单一、导师主导创意、学生专业同质性高。首先，在课堂成果形式单一方面，目前与作业挂钩的设计成果过于抽象且滞后，往往无法满足瞬息万变的职场对人才的期望，导致培养出来的学生与企业需求存在较大差距；在导师主导创意方面，目前课堂作业的产出方式大多以导师为中心，这使得学生们对教师过度依赖，阻碍了他们的自我反思和设计思考，导致学生的设计创新能力不足；在学生专业同质性较高方面，由于同班同学之间的专业划分细致且同质性高，在课堂组队设计过程中缺乏与其他专业同学之间的互动，导致学生在学习过程中的积极性和行动力受限。

In 2009, Harvard University Education Fellow Chris Dede proposed "Skills for the 21st Century", emphasizing that the breadth of "skills" needs to be composed of skills, abilities, and dispositions to learn. Thus, it is also clear that these "skills" are the various ways in which students acquire and master knowledge and technology in a given context. These "skills" can be divided into three types: perceptual, operational, and applied, which have a high level of intelligence and adaptability to work and can perform their tasks effectively and respond positively to others. Over the years, we can see from the changes in college admissions that schools are increasingly demanding a diversity of professional skills in the selection of designers and that design instructional planning for students after they enter the school not only needs to continue to enhance students' "T-shaped" horizontal skills but also places great emphasis on students' ability to work collaboratively, to shift the focus from the "individual" learning environment of the past. The transition from an "individual" learning environment to a collaborative, "team" environment is critical.

However, even though the current design teaching model attempts to provide a multi-disciplinary and diversified training vision, our design classrooms still have several pain points, namely, single classroom outcomes, tutor-driven creativity, and high homogeneity of students. In terms of the single form of classroom results, the current design results linked with assignments are abstract and lagging, often failing to meet the expectations of the rapidly changing workplace for talents, resulting in a large gap between the cultivated students and the needs of enterprises; in terms of tutor-led creativity, the current output of classroom assignments is mostly tutor-centered, which makes students more dependent on teachers and hinders self-reflection. In terms of homogeneous students in the class, due to the high homogeneity of majors among students in the same class, team design process often lacks interaction with students from other majors or other schools, resulting in limited motivation and action of students in the learning process.

此时，我们需要做的就是正视上述痛点，以更积极的心态去改革现有的设计教学模式。在进行"创意与实践"课程时，要根据学生实际情况制定合理合适的教学方法，对教学过程中使用到的相关材料进行分析总结，使其更符合学生的认知规律，为后续学习奠定良好基础。尤其是在指导专业方向各异的设计学生中常能发现，他们如出一辙地更想体验新奇的、有创意的学习环境。对于"学设计"这种充满时代创造性的教学活动，采用适应当前年轻学生们的新形态教学形式更能激发学生们的学习动力。

设计工作坊在艺术类学生培养计划中属于"实验与设计工作坊课程"的范畴，此类型课程的本质是深度学习，意味着学生们不仅要"知道"还要能"做到"。工作坊体验式的教学方法是以"促进学生发现学习"为主的一种教学方式，是在教师有目的、有计划的引导下，让学生们自己更主动体验知识从而理解知识的途径。作为诸多实践教学模式之一，设计工作坊不仅能帮助学生有效地将知识应用到实践当中从而积累经验，还能让老师有更多机会进行课程研究并使课堂更具互动性，这种形式的实践性课程能在充分利用社会与行业资源的同时，将其快节奏与参与度高的特点充分应用在学生学习积极性的激发上。而设计马拉松就是基于以上传统课堂痛点所产生的设计工作坊，作为满足职场快节奏步调所安排的短期课程，设计马拉松为未来设计师候选人提供了在不同学科领域获得能力与经验的机会，它要比纸上谈兵的传统设计教学方式有更高的学习多样性与产出要求。

作为北京设计周期间代表性的创新学术活动之一，设计马拉松自2018年开始与世界各知名高校联合探索线下线上混合式学习环境的教学手段，为此将工作坊的前期课程内容转移到线上进行，于此可允许更多国际专家和学员们进行远程交流，让原本无法参加线下活动的人员也能在其所在地轻松参与丰富的课程活动。

从时间维度来看，线下线上混合式工作坊依托于互联网环境，老师能迅速有效地随时开展教学活动，包括课前预习、资源发布、课后辅导等。学生们的学习时间可以根据不同需求随时调整，不会局限于有限的工作坊授课范围之内。丰富且高效的线上协作工具能更好地让全球学员在不同时区的地理条件限制下协同工作，并帮助不同语种的设计师快速进行沟通。

We need to face the above pain points and reform the existing design teaching mode with a more positive attitude. When conducting the "Creativity and Practice" course, we should develop reasonable and appropriate teaching methods according to the actual situation of students, analyze and summarize the relevant materials used in the teaching process, to make them more consistent with the characteristics of students' cognitive rules and lay a good foundation for subsequent learning. We often find that design students with different directions want to experience a new and creative learning environment. The new forms of teaching and learning, which are adapted to the current generation of young students, are an excellent motivation for students to learn design.

Design workshops are part of the "Experimental and Design Workshop Program" in the art students training program. The essence of this type of program is deep learning, which means that students must not only "know" but also "do". The experiential workshop approach is a way to promote discovery learning, a method for students to experience and understand knowledge on their initiative, with the teacher's guidance in a purposeful and planned way. As one of the many practical teaching modes, design workshops not only help students effectively apply their knowledge to practice and gain experience but also provide teachers with more opportunities to conduct research and make the classroom more interactive. This form of practical course can make full use of social and industry resources while applying its fast-paced and participatory characteristics to motivate students. The Design Day Marathon is a design workshop based on these traditional classroom pain points. As a short course arranged to meet the fast pace of the workplace, the Design Day Marathon provides future designer candidates with the opportunity to gain competence and experience in different disciplines, and it has a higher level of learning diversity and output requirements than traditional design teaching on paper.

As one of the representative and innovative academic events during Beijing Design Week, Design Day Marathon has been exploring offline and online blended learning environments with world-renowned universities since 2018, by transferring the pre-course content of the workshops online, allowing more international experts and participants to communicate remotely and those who would otherwise not be able to attend offline events to participate in the rich course activities at their location easily.

Regarding the time dimension, the offline and online hybrid workshops are based on the Internet environment, allowing teachers to conduct teaching, and learning activities quickly and effectively at any time, including pre-learning, resource distribution, and post-learning tutorials. Students' learning time can be adjusted according to different needs and will not be limited to the limited scope of the workshop. The rich and efficient online collaboration tools better enable global students to work together under the geographical constraints of different time zones and help designers of other languages to communicate quickly.

从课程体验来看，设计马拉松通过"体验式教学"来体现教师的导引、学生的参与和师生的交流这三种类型的互动方式。在挖掘设计痛点的过程中引导学生去找到多重解决方案，以互动性强和灵活性大的课程安排满足不同院校之间学员们的需求。在实施过程中，体验式教学手段更是一种效力持久的教学方式，它能集知识（knowledge）、实践（activity）和反馈（reflection）于一体，以"行动"为中心促进学生学习，并且让他们多方面地反馈学习过程与成果。

从空间维度来看，组织国际性设计教学可以发展学生跨文化学习的能力，同时也可以增加他们对于本土文化的认同以及对于其他文化的了解与包容。在工作坊的执行过程中让学生以自信的态度面对跨文化交流产生的意见分歧，在加强其专业设计能力的同时也让学员们适当理解不同文化的价值。通过对跨文化学习能力的培养，学生们能有效地接触到各种文化背景的专业设计师们，并较好地在这个越来越多样化的国际环境中学习和生活，以充足的经验去观察和处理复杂的设计问题。

设计马拉松的出现，无疑是为当前的设计教学模式打开了一扇门，它构建了一个能促进学生实践性思维并且不断发展创新的辅助教学活动，基于企业与社会所需的实践性问题开展体验式教学，并对其有效性进行落地验证。有序且合理地反思当前艺术设计教学体系中关于教学目的、教学过程、教学原则、教学方法等实施过程的合理性，分析有关教学改革的优势与局限性，探寻更为和谐的师生关系，提出新颖的课程和教学策略。在此基础上建构体验式教学范式，从设计教育的未来发展出发，从提升学生参与度的方向切入，去探究如何利用协作工具加强线上线下融合教学体验，培养学生在跨文化交流过程中的包容性，进而有效提升设计师们在院校中培养的综合专业素质。

Regarding the course experience, Design Day Marathon embodies three types of interaction through "experiential teaching": teacher guidance, student participation, and student-teacher communication. In uncovering design pain points, students are guided to find various feasible solutions to meet the needs of students from different institutions with a highly interactive and flexible curriculum. In its implementation, the experiential teaching method is an even more effective and long-lasting way that integrates knowledge, activity, and reflection to facilitate students' learning with "action" as the center, allowing them to give feedback on the learning process and results in many ways. Process and results.

From a spatial perspective, organizing international design instruction develops students' ability to learn cross-culturally, while increasing their cultural identity and understanding and tolerance of other cultures. In the process of implementing the workshop, students can face the differences of opinion in the process of intercultural communication with confidence, strengthening their professional design skills and giving them a proper understanding of the value of different cultures. Through the development of cross-cultural learning skills, students can effectively reach out to professional designers from various cultures and better learn and live in this increasingly diverse international environment with sufficient experience to observe and deal with complex design issues.

The emergence of the Design Day Marathon has undoubtedly opened the door to the current design teaching model, building a supplementary teaching activity that promotes practical thinking and continuous development of innovation among students, conducting experiential teaching based on the practical problem-solving required by enterprises and society, and verifying its effectiveness on the ground. We will reflect on the rationality of the current art and design teaching system in terms of teaching purpose, teaching process, teaching principles, and teaching methods, analyze the major advantages and limitations of the teaching reform, explore a more appropriate teacher-student relationship, and propose a novel curriculum and teaching strategy. Based on this, we will construct an experiential teaching paradigm, start from the future development of design education and enhance student participation, explore how to use collaborative tools to strengthen online and offline integration teaching experience, promote students' inclusiveness in the process of cross-cultural communication, and then effectively improve the comprehensive professional quality of designers in the process of institutional training.

丁肇辰教授
设计马拉松创始人

Professor Chawchen Ting
Founder of The Design Day Marathon

每当我画下一个圆圈，立马我就想走出这个圆圈。

理查德·巴克敏斯特·富勒

Whenever I draw a circle, I immediately want to step out of it.

Richard Buckminster Fuller

目录

01

跨文化的设计实践平台

03

2021 设计马拉松

02

2022 设计马拉松

04

致谢

CONTENTS

01

跨文化的设计实践平台

INTERCULTURAL DESIGN
PRACTICE PLATFORM

1.1 设计马拉松的目的何在

1.1 WHAT IS THE PURPOSE OF THE DESIGN DAY MARATHON

1.1.1　快速获取综合性设计知识

设计马拉松因其能迅速促进学生学习设计的效果而备受好评。它能以关注当前社会的快速变化和重要需求为出发点，使传统的设计教学模式发生质的转变。此外，设计马拉松的活动内容具备多样性，每个部分都能为学生提供不同的学习时间和学习目标。以即兴直播间为例，它为国内外的学生提供了快速学习设计课题和了解新技术的机会，它的优势在于能够提供不同设计专业的动态信息，这使学生能够在很短的时间内获取综合性的知识，也使多国之间的沟通更加便捷。

1.1.2　互联网速度下的快速课题产出

设计马拉松工作坊与一般设计课程不同之处在于，它容许学生在很短的时间内快速地实现设计产出。在传统的设计课程中，需要近四个月的周期，期间包含冗长的设计调研、设计草案、用户研究及设计测试等。如今在互联网时代，传统设计课程的安排会让学生感觉不到太多来自时间的压力，导致其设计效率低下、产出结果缓慢。然而，设计马拉松能够适当地调整此情况，帮助学生节省大量的设计实践时间，在短期内取得相对完善的成果并达到学习目标。

1.1.3　激发国际化学习氛围下的竞争力

在工作坊中，跨国导师们以丰富的教学经验引导学生们朝着正确的设计方向前进，并激发他们的创新思维。这些来自国内外不同高校的学生们通过合作和互相学习，希望产出更优秀的课程成果。同学们在分组后逐渐形成了一种良性的竞争环境，他们相互比较、互相超越，打破了学校的传统教学模式，也开拓了新的学习领域。这种竞争态势让学生们拥有更强烈的目标感，使他们更加努力地追求卓越，这种跨院校的学习模式让学生们在竞争中成长，在合作中取得进步，为他们未来的学习打下了坚实的基础。

1.1.1　Quickly acquire comprehensive design knowledge

The Design Day Marathon is highly acclaimed for its ability to quickly facilitate students' design learning. It focuses on the rapid changes and essential needs of the current society as a starting point and makes a qualitative transformation of the traditional design teaching model. In addition, the activities of the Design Day Marathon are diversified, and each part can provide students with different learning times and objectives. Taking Live Jam as an example, it provides students at home and abroad with the opportunity to quickly learn design topics and understand new technologies, and its advantage is that it allows for dynamic information about different design specialties, which enables students to acquire comprehensive knowledge in a short period. It also makes the communication between multiple countries more convenient.

1.1.2　Rapid subject output at the speed of the Internet

The Design Day Marathon workshop is different from a typical design course in that it allows students to quickly achieve design outputs in a very short period. In a traditional design program, it takes nearly four months of lengthy design research, design drafts, user research, and design testing. Nowadays, in the age of the Internet, traditional design courses are organized in such a way that students do not feel much time pressure, which leads to inefficiency in their design and slow output. However, The Design Day Marathon can adjust this situation and help students save time for design practice, achieve relatively perfect results, and meet the learning objectives in a short time.

1.1.3　Stimulating Competitiveness in An Internationalized Learning Environment

During the workshop, the multinational instructors with rich teaching experience guided the students in the right direction of design and inspired them to think creatively. The students from different universities in China and abroad collaborated and learned from each other, hoping to achieve better course outcomes. The grouping of students creates a healthy competitive environment in which they compare and surpass each other, breaking the traditional teaching model of the school and opening new areas of learning. This competitive dynamic gave the students a stronger sense of purpose and made them work harder to pursue excellence. This inter-institutional learning model allowed the students to grow through competition and progress through cooperation, laying a solid foundation for their future studies.

1.2 设计马拉松与一般工作坊之间有何不同

1.2 WHAT IS THE DIFFERENCE BETWEEN A DESIGN DAY MARATHON AND A WORKSHOP

工作坊是设计马拉松最重要的环节，每年我们邀请全球导师基于该年主题发布设计课题，学员可在注册后选定相应导师加入课题组。和一般工作坊不同之处在于，设计马拉松有着明显的竞赛性质，学员来自国内外院校，造就了参赛组员的多样性。设计马拉松工作坊主要包括"报名与作品集筛选"和"线上工作坊与答辩"两个阶段。在前一阶段，组委会筛选作品集，通过作品集筛选的学员则可加入马拉松工作坊；未通过作品集筛选的学员可自行组队，参与中期评选。在后一阶段，线上工作坊由各导师主持并把握指导时间，中期检查通过线上方式汇报方案，并由导师组进行评分，挑选出15组晋级到最终答辩，入选小组将继续推动项目的落地直到最终答辩之日（组委会视招生情况进行调整）。

1.2.1 具备落地性的研究成果

"工作坊"式的设计教育具备较长历史，其源头可以追溯到20世纪初。当时的设计教育工作者们倡导将手工艺与艺术相结合，以工作坊的形式进行实践教学。这种教育模式着重培养学生的实践操作能力和创新思维，同时注重团队协作和跨学科交流。设计马拉松工作坊的特点在于通过线上与线下的同步执行，不仅延长了一般工作坊的授课时长，使学生更深入、细致地参与设计的过程。此外，来自不同国家地区的学员们组队合作，专注于产出能真实落地的成果。作为工作坊的变体，设计马拉松工作坊具有创新性、实用性、跨学科性和可持续性等特征，对于推动设计教育的发展和进步具有积极的作用。这种教育模式的探索，在培养新一代设计师方面具有重要意义。

The workshop is the most critical part of the Design Day Marathon. Every year, we invite global tutors to issue designs based on the year's theme. Students can choose the corresponding tutors to join the task groups after registration. The difference from general workshops is that the Design Day Marathon has an obvious nature of the competition, and the students come from domestic and foreign universities, which creates a diversity of participating team members. The workshop includes two parts: registration & portfolio review and online workshop & reviews. In the former part, organizing committees will review the portfolio, and the students whose works are selected will be enrolled. Students who were not selected can team up with each other and participate in the mid-term inspection. In the latter part, the tutors of each group with flexible schedules will host the online workshop. After the intermediate review, the tutors will score for each group. Only fifteen groups can continue to finish their project and participate in the final defense (The organizing committees will adjust according to the enrollment situation).

1.2.1 Research Results That Are Grounded

Design education in the form of workshops has a long history, dating back to the early 1900s. At that time, design educators advocated the combination of crafts and art, and practical teaching in the form of workshops. This mode of education emphasizes the development of practical skills and creative thinking, as well as teamwork and interdisciplinary communication. The Design Day Marathon workshop is characterized by the synchronized execution of online and offline, which not only extends the length of a typical workshop's instruction but also enables students to participate in the design process in a more in-depth and detailed manner. In addition, students from different countries and regions team up and collaborate, focusing on producing results that can be realistically implemented. As a variation of the workshop, the Design Day Marathon workshop is characterized by innovation, practicality, interdisciplinary, and sustainability, and has a positive effect on promoting the development and progress of design education. The exploration of this educational model is of great significance in cultivating a new generation of designers.

1.2.2　跨国与跨文化的交流学习契机

学生们在国内和国外学习设计的方法和手段本身存在差异，这需要导师们具备跨文化交流的能力和对不同设计理念的敏感度。设计马拉松为学生们提供了一个较为合适的国际化交流机会，使学生们能够更快地接触国际化的设计方法。在指导过程中，国际导师们不仅提供了丰富的资源和工具，还鼓励学生之间的合作和分享，让学生在国际化交流中获得更多的机会和成长。同时，导师们也需要适当地了解国内的设计环境，以及学生们的设计态度和方法习惯等。在此学习交流中，学生们不仅可以获得更广阔的视野和更多的机会，还可以通过与不同文化背景的学生和导师的互动，培养自己的跨文化交流能力和团队合作精神，为未来的职业生涯奠定基础。

1.2.3　面向企业与社会需求的设计课题

实践性教育是培养设计学科人才的关键，有助于学生了解实际设计问题并提升他们的敏感度和灵活性。这种学习方式能够帮助学生适应行业需求，提高他们在未来就业时的竞争力。实践性教育的另一个重要方面是关注全球热点和发展动态。在培养优秀设计师的过程中，应该让他们时刻关注宏观的发展需求和热点议题，以便能够将优秀的设计理念和经验融入自身的设计。设计马拉松是一种创新的教育方式，通过实践性教育、关注社会热点和发展动态，通过全球化的协同设计行动，可以使设计实践的思考更加全面，在重大设计课题的解决方案中做出更大的贡献。

1.2.2　Transnational and Intercultural Learning Opportunities

There are inherent differences in the methods and means by which students learn design at home and abroad, which require tutors to have cross-cultural communication skills and sensitivity to different design concepts. The Design Day Marathon provides students with a more suitable opportunity for international exchange, enabling them to get in touch with international design methods more quickly. During the tutoring process, the international tutors not only provide abundant resources and tools but also encourage cooperation and sharing among students, so that they can gain more opportunities and growth in international exchange. At the same time, the tutors also need to properly understand the domestic design environment, as well as the students' design attitudes and methodological habits. In this learning exchange, students can not only get a broader vision and more opportunities but also develop their cross-cultural communication skills and teamwork spirit through interaction with students and tutors from different cultural backgrounds, to prepare for their future careers.

1.2.3　Design Subjects Oriented to the Needs of Enterprises and Society

Practical education is the key to cultivating talents in design disciplines, helping students to understand actual design problems, and enhancing their sensitivity and flexibility. This learning approach helps students to adapt to the needs of the industry and enhances their competitiveness in future employment. Another important aspect of practical education is to pay attention to global hotspots and developments. In the process of cultivating outstanding designers, they should be made to keep an eye on macro development needs and hot topics so that they can incorporate excellent design concepts and experiences into their designs. The Design Day Marathon is an innovative way of education., where by practical education and attention to the hot topics and dynamics of society, as well as globalised collaborative design practice and a better contribution to the solution of major design issues.

02

2022 设计马拉松
2022 DESIGN DAY MARATHON

2.1　2022 设计马拉松总体安排

2.1　2022 DESIGN DAY MARATHON GENERAL ARRANGEMENTS

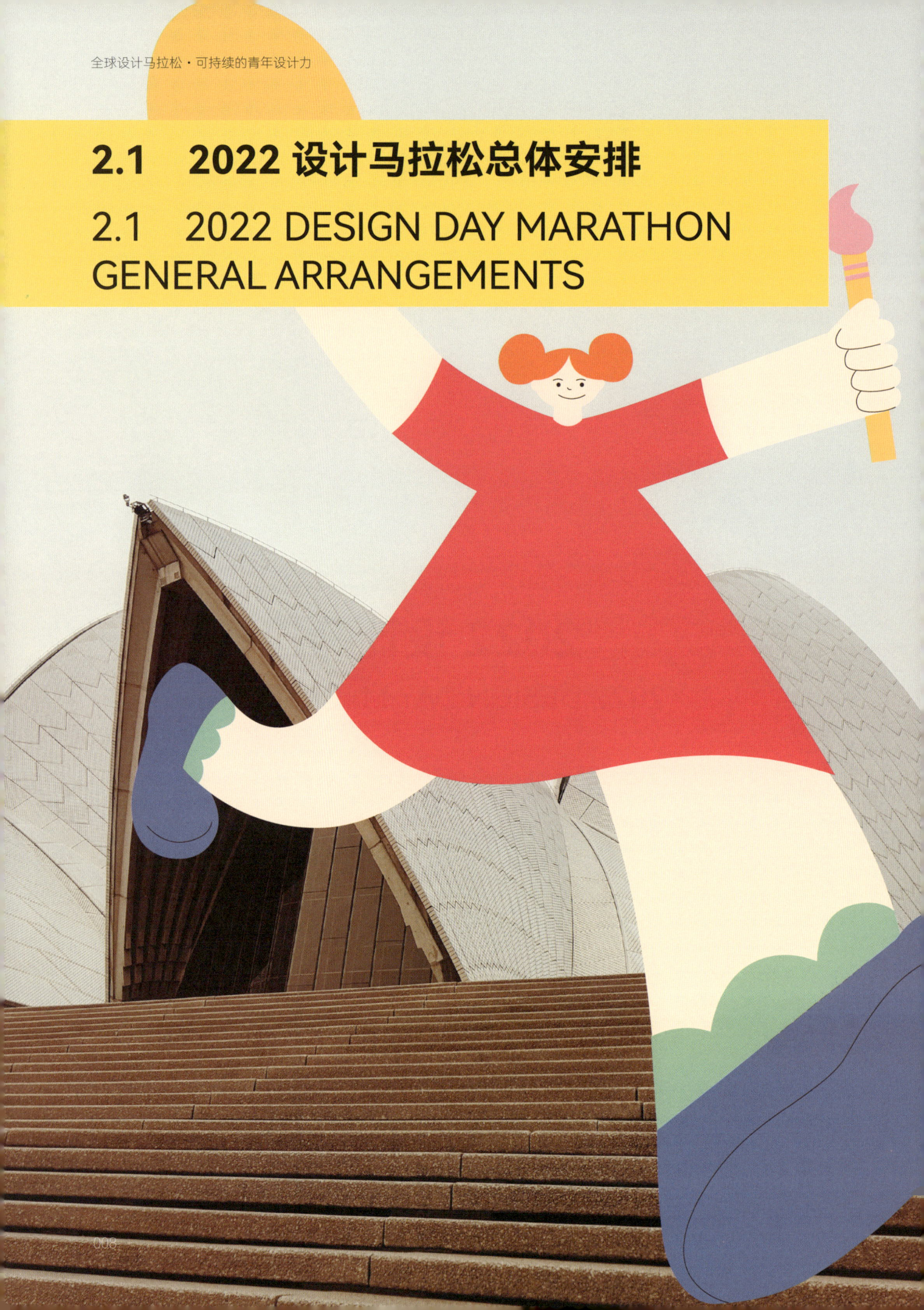

2.1.1　主题：优雅的生活

世界正在经历前所未有的重置，我们的生活变得不那么"优雅"。但令人庆幸的是，随着这个"重置"的进行，人们对优雅生活的渴望变得更加强烈，我们开始看到一个崭新时代的来临，有望让人们的生活变得更美好。

"优雅"并不是指物质上的奢侈，也不是对名利的追求，更不需要大量的追随者。优雅可以体现在更深层次的人际关系中，可以是对未来持开放和包容的态度，可以是更强烈的自我满足感，也可以是对生活更深层的欣赏。

2022 年的设计马拉松让我们拥有开阔的胸怀，懂得取舍，谦虚有礼地对待他人，关注地球危机，为社会大众做出贡献，照顾好朋友和家人。让我们从被打乱的生活中发现更优雅的生活态度，一起成为优雅的生活者。

2.1.1　THEME: GRACEFUL LIFE

The world is going through an unprecedented reset, making our lives graceless. Fortunately, with the "reset", people's desire to live a graceful life has grown stronger, and we're starting to see an era of rebirth taking shape that promises to make people's lives better.

"Graceful" doesn't mean material luxury, nor is it the support of fame and wealth, nor does it need a lot of people to pursue. "Graceful" can mean deeper interpersonal relationships, an open and inclusive attitude towards the future, a stronger sense of self-satisfaction, or a greater appreciation of life.

The 2022 Design Day Marathon allows us to keep a broader mind, know when to hold fast and let go, treat others with humility and courtesy, pay attention to the global crisis, contribute to the public, and love our friends and family. Let's find a more graceful attitude to the life disrupted and live graceful lives together.

2.1.2　2022 设计马拉松统计数据

2.1.2　2022 DESIGN DAY MARATHON STATISTICS

46	126	12
参与地区数量 Countries & Regions	参与高校数量 Schools	参与专家数量 Experts
345	60	59
工作坊学员数量 Workshop Participants	工作坊导师数量 Workshop Tutors	工作坊方案数量 Workshop Projects
35	20	9
支持学校数量 Supporting Schools	支持媒体数量 Support Media	学术委员数量 Academic Committees
21	12	500000
组委会数量 Organizing Committees	直播间数量 Live Streamings	直播观看数量 Live Streaming Views

2.1.3　2022 设计马拉松日程　2.1.3　2022 DESIGN DAY MARATHON SCHEDULE

策划期
Preparation

宣讲期
Announcement

报名期
Pre-Registration

2—4 月 / February-April

活动策划
合作洽谈

Activity Planning
Cooperation Negotiation

4—6 月 / April-June

院校线上宣讲
邀请导师

Institutional Online Presentations
Inviting Tutors

6—8 月 / June-August

工作坊答疑会
学员报名筛选
确认学员名单

Workshop Q&A
Student Pre-registration
Announcing Registration Results

即兴直播间
Live Jam

工作坊
Workshop

9 月 / September

主题直播间
工作坊直播间

Theme Live Jam
Workshop Live Jam

9 月 / September

学员进入课题组
坊前答疑"闪电营"
线上工作坊开始
线上中期检查
线上最终答辩

Students Entering Task Groups
Pre-workshop Q&A "Lightning Camp"
Start of Online Workshop
Online Intermediate Inspection
Online Final Defense

2.1.4　即兴直播间

即兴直播间是线上主题沙龙，分为线上圆桌会议与线上课题宣讲两部分，这是学员在进入课题组之前认识课题与熟悉导师的最佳渠道。

2.1.4　LIVE JAM

Live Jam is an online theme salon, which is divided into two parts: an Online Theme Forum and an Online Tutors Roundtable. It is the best channel for students to get to know the subjects and get familiar with the tutors before participating in the task groups.

线上主题论坛 / Online Theme Forum　　　9 月 17 日 / September 17th

10：00—11：00 设计教育学生说：清华+同济+米兰理工（"新常态"下的人居环境）（杨叶秋，意大利米兰理工大学）
10：00-11：00 Design Education Students Say: Tsinghua+Tongji+Miri (Residential Environment under the "New Normal") (Yeqiu Yang, Polytechnic University of Milan, Italy)

11：00—12：00 传统设计新生力（刘愿，云南艺术学院）
11：00-12：00 New Forces in Traditional Design (Yuan Liu, Yunnan Arts University)

13：00—14：00 城市用户体验设计（潘荣焕，韩国国民大学）
13：00-14：00 City UX Design (Younghwan Pan, Kookmin University, Korea)

14：00—15：00 波希米亚人再现（Hambalee Jahma，泰国宋卡王子大学）
14：00-15：00 Bohemians Re-creation （Hambalee Jahma, Prince of Songkla University, Thailand）

16：00—17：00 设想未来——设计教育国际驱动力（Martin Liu，中欧国际设计文化协会）
16：00-17：00 Envisioning the Future—International Driver for Design Education (Martin Liu, China Europe International Design Culture Association)

17：00—18：00 优雅的聆听——《流动的音诗》中国当代作曲家专场音乐会（亓梦婕，中国音乐学院）
17：00-18：00 Elegant Listening—"Flowing Poetics" A Concert of Contemporary Chinese Composers (Mengjie Qi, China Conservatory of Music)

线上导师圆桌会议 / Online Tutors Roundtable　　　9 月 18 日 / September 18th

10：00—11：00 有组织的声音——致敬埃德加·瓦雷兹（Edgard Varèse）专场电子音乐会（亓梦婕，中国音乐学院）
10：00-11：00 The Organized Sound—A Tribute Concert to Edgard Varèse (Mengjie Qi, China Conservatory of Music)

11：00—12：00 一起来设计吃！（何颂飞，北京服装学院）
11：00-12：00 Let's Design Eating! (Songfei He, Beijing Institute of Fashion Technology)

13：00—14：00 下一步计划：洞察你的未来发展与职业倾向（丁肇辰，北京服装学院）
13：00-14：00 What's Next: Insight into Your Future Development and Career Orientation (Chawchen Ting, Beijing Institute of Fashion Technology)

14：00—15：00 虚拟时尚拟态设计（魏勤文，北京服装学院）
14：00-15：00 Virtual Fashion and Mimetic Design (Qinwen Wei, Beijing Institute of Fashion Technology)

15：00—16：00 现在和过去的设计师（Hanny Wijaya，印度尼西亚建国大学）
15：00-16：00 Designers Now and Then （Hanny Wijaya, BINUS University, Indonesia）

16：00—17：00 感官体验：真实与虚拟之间（Anne Farren，澳大利亚科廷大学）
16：00-17：00 Sensory Experiences: Between the Real and Virtual （Anne Farren, Curtin University, Australia）

2.1.5 工作坊课题

2022年设计马拉松工作坊课题、院校、方向（排名不分先后）：

2.1.5 WORKSHOP TASKS

2022 Design Day Marathon workshop tasks, institutions, and directions (in no particular order):

	课题	院校：方向
1	更优雅的生活 A More Elegant Life Era	（中央美术学院：产品设计） (Central Academy of Fine Arts: Product Design)
2	如何优雅地运动 How To Exercise Gracefully	（中原大学：交互设计、产品设计） (Chung Yuan Christian University: Interaction Design, Product Design)
3	我们开始吧——优雅的行为 Let's Do It—Graceful Behavior	（南京信息工程大学：交互设计、环境设计） (Nanjing University of Information Science And Technology: Interaction Design, Environmental Design)
4	老龄化社会背景下的公共产品设计 Public Product Design In The Context of An Aging Society	（南京艺术学院：交互设计、产品设计） (Nanjing Arts Institute: Interaction Design, Product Design)
5	老龄社区服务系统设计 Aging Community Service System Design	（北京邮电大学：服务设计） (Beijing University of Posts And Telecommunications: Service Design)
6	如何优雅地时尚——西南非遗的数字化时尚设计 How To Maintain Fashion Gracefully—Digital Fashion Design of Southwest Intangible Cultural Heritage	（四川大学：视觉传达、时尚设计） (Sichuan University: Communication Design, Fashion Design)
7	乡村持续力——艺术设计赋能乡村振兴 Task: Rural Sustainability—Art & Design Empowers Rural Revitalization	（鲁迅美术学院：服务设计、产品设计） (Luxun Academy of Fine Arts: Service Design, Product Design)
8	虚拟时尚拟态设计 Virtual Fashion Mimetic Design	（北京服装学院：服装与服饰设计、产品设计） (Beijing Institute of Fashion Technology: Fashion And Apparel Design, Product Design)
9	幸福感成长桌游 Growing Happiness Board Game	（北京服装学院：游戏化设计、产品设计） (Beijing Fashion Institute of Technology: Gamification Design, Product Design)
10	小确幸大生活——定格动画设计课程 Small Fortunes, Big Lives—Stop-Motion Animation Design Course	（天津美术学院：动画设计） (Tianjin Academy of Fine Arts: Animation Design)
11	以身体之——具身性设计课程 Designing With The Body—Embodied Design Course	（中国美术学院：时尚设计） (China Academy of Art: Fashion Design)
12	增进获得：为亚洲用户的无障碍日常生活而设计 Enhancing Access: Designed For The Accessible Everyday Life of Asian Users	（北京服装学院：无障碍设计、包容性设计） (Beijing Institute of Fashion Technology: Accessible Design, Inclusive Design)
13	15分钟生活圈："新常态"下的人居环境设计 15-Minute Living Circle: Residential Environment Design Under The "New Normal"	（意大利米兰理工大学：服务设计、环境设计） (Polytechnic University of Milan, Italy: Service Design, Environmental Design)
14	优雅地退休——包容性设计赋能老龄产品创新 Retire Gracefully—Inclusive Design Empowers Innovation In Aging Products	（北京航空航天大学：服务设计、产品设计） (Beihang University: Service Design, Product Design)
15	声音的感知与重构 Sound Perception And Reconstruction	（中国音乐学院：音乐设计、声音可视化） (China Conservatory of Music: Music Design, Sound Visualization)
16	购买本土产品——图形标识 Buy Local—Graphic Identity	（葡萄牙马托西纽什艺术与设计学院：概念设计、环境设计） (Esad Matosinhos, Portugal: Conceptual Design, Environmental Design)
17	设计师宣言：内容即信息 Designers Manifesto: The Content Is The Message	（英国新白金汉大学：设计哲学、设计宣言） (The New University of Buckingham, Uk: Design Philosophy, Design Manifesto)
18	设计向善 Design For Good	（新加坡南洋理工大学：平面设计、伦理设计） (Nanyang Technological University, Singapore: Graphic Design, Ethical Design)
19	数字时代的数字时尚与数字传播 Digital Fashion And Communication In Digital Age	（韩国祥明大学：数字时尚、3D数字创作） (Sangmyung University, Korea: Digital Fashion, 3D Digital Creation)
20	通过九型人格特征进行自我发现 Self Discovery Through Enneagram Branding	（美国阿肯色大学：字体设计、品牌设计） (University of Arkansas, USA: Typography, Branding)
21	通过无声之物：重塑奥地利木偶表演 Through The Silent Objects: Re-Discover The Manifestation of Austrian Puppetry	（奥地利林茨艺术大学：电影动画、产品设计、环境设计） (University of The Arts Linz, Austria: Film Animation, Product Design, Environmental Design)
22	优雅生活烹饪书 Graceful Life Cookbook	（英国密德萨斯大学：平面设计、插画设计） (Middlesex University, UK: Graphic Design, Illustration Design)
23	有意义的设计：在虚拟世界中设计的"优雅生活"的构建 Meaningful Design: The Construction of A Graceful Life Designed In A Virtual World	（澳大利亚科廷大学：电影动画、交互设计） (Curtin University, Australia: Film Animation, Interaction Design)

2.1.6　工作坊预期产出

2.1.6　WORKSHOP EXPECTED OUTPUTS

生活方式设计 Lifestyle Design	视觉传达设计 Communication Design	概念设计 Conceptual Design
APP 设计 APP Design	视频短片 Short Video	未来设计 Future Design
产品设计 Product Design	动画设计 Animation Design	服务设计 Service Design
网站设计 Web Design	虚拟形象 IP Design	广告策划 Advertising Campaign
环境设计 Environmental Design	互动艺术 Interactive Art	公关策划 PR Proposal
时尚设计 Fashion Design	互动广告 Interactive Advertising	活动设计 Event Design
文创产品 Creative Products	虚拟时尚 Visual Fashion	

2.2　2022 设计马拉松工作坊成果

2.2　2022 DESIGN DAY MARATHON WORKSHOP RESULTS

↘ **2.2.1 课题：更优雅的生活**

2.2.1 TASK: A MORE ELEGANT LIFE ERA

课题说明：

在经历了担忧、感伤、焦虑及对未来的不确定感之后，原有的生活秩序被打破，我们的生活变得不优雅。如何通过设计的力量，让生活变得更加优雅，是一个值得思考的问题。设计，先天具有交叉学科或跨学科的性质。现代设计越来越多地具备文化意义，并与美学、文学、工程学和科学技术等融合。课题将以家居产品为载体，结合文化、美学、感官设计理论等，以艺术手段结合五感进行家居产品设计，以提升人在家居环境中的生理及心理舒适度，更加优雅地生活，感受生活的美好。

指导老师：

陈娜，中央美术学院城市设计学院教师

Task Description:

After experiencing worry, sadness, anxiety, and uncertainty about the future, it breaks the original life order and makes our life graceless. How to make life more graceful through the power of design is a question worth considering. The design has an innate interdisciplinary nature. Modern design increasingly has the meaning of culture, aesthetics, literature, engineering science and technology, etc. This project will take home products as the carrier, combining culture, aesthetics, and sensory design theory. It uses artistic means to combine the five senses to design home products, to improve people's physical and psychological comfort in the home environment, to live a more graceful life, and to feel the beauty of life.

Tutor:

Na Chen, Lecturer of the School of Urban Design, Central Academy of Fine Arts

2.2.1.1　绿界——室内蔬菜种植装置

作者：郝伟赓、郭承沂、林碧珊、王子涵、张昊哲

　　该装置主要由水循环箱体搭配若干模块化种植箱构成，不同的绿界用户可根据家庭实际情况自由配置箱体数量，也可以根据家庭实际情况选择箱体摆放位置。外部的立方体灯架，既是水培植物的光补偿装置，在室内灯光不佳时，也可作为室内的氛围灯使用。绿界既是家庭的蔬菜生产者，可以为家庭提供一定的蔬菜供应，也是一个富有情调的情绪调节者，它的气味散发装置可以散发出植物香气，疗愈心灵，并且种植本身也是一种减压治愈、维护家庭关系的良好方式。绿界的主要灵感来自"蜂巢"。蜂巢造型是物质世界中较稳定的几种结构之一，蜜蜂在蜂巢中各司其职，形成比较稳定的微型社会，绿界也希望在家庭里营造出相互理解，相互包容的和谐关系，就和绿界的名字一样，代表着安全和放心。

2.2.1.1　Green Zone—Indoor Planting Device

By Weigeng Hao, Chengyi Guo, Bishan Lin, Zihan Wang, Haozhe Zhang

　　The device is mainly composed of a water circulation box and several modular planting boxes. Different Green Zone users can freely configure the number of boxes according to the actual situation of the family or choose the location of boxes according to the actual situation of the family. The external cube lamp holder is not only a light compensation device for hydroponic plants but also can be used as an indoor atmosphere lamp when the indoor light is poor. The Green Zone is a vegetable producer in the family and a sentimental mood regulator. Its smell-emitting device can emit plant fragrance, and heal the mind, and planting itself is also an excellent way to reduce pressure, cure, and maintain family relations. The main inspiration for the Green Zone comes from the "beehive". Because the shape of the beehive is one of the more stable structures in the material world, bees perform their duties in the beehive to form a more stable micro-society. The Green Zone also hopes to create a harmonious relationship of mutual understanding and mutual tolerance in families. Just like the name of the Green Zone represents safety and reassurance.

2.2.1.2　光愈

2.2.1.2　Light Healing

作者：张金渠、毛咸、谢千慧、杨泽贤、闫怡霏

By Jinqu Zhang, Xian Mao, Qianhui Xie, Zexian Yang, Yifei Yan

　　有一类人因为种种原因不愿意走出家门与人社交。我们试图探知这类人群的居家生活状态，剖析他们细腻敏感的情绪色彩，为其提供珍贵的情绪体验与情绪价值。本设计以纤维艺术为载体，运用柔性金属材料，最终产出了一组分为两个系列的灯具作品。希望通过该组作品呈现"居家后遗症人群"脆弱敏感的内心状态，引发人们的共情与关心，并希望以温柔的光线来治愈他们脆弱敏感的居家时光。

　　For various reasons, a group of people are reluctant to go out of their homes to socialize with people. We try to explore the state of their home lives, dissect their delicate and sensitive emotions, and provide them with precious emotional experience and emotional value. This design uses fiber art as a carrier and flexible metal to produce a group of lamps divided into two series. It is hoped that through this group of works, the fragile and sensitive inner state of the "home sequelae of the people" will be presented to arouse people's sympathy and concern, and the fragile and gentle light will be used to heal their fragile and sensitive home time.

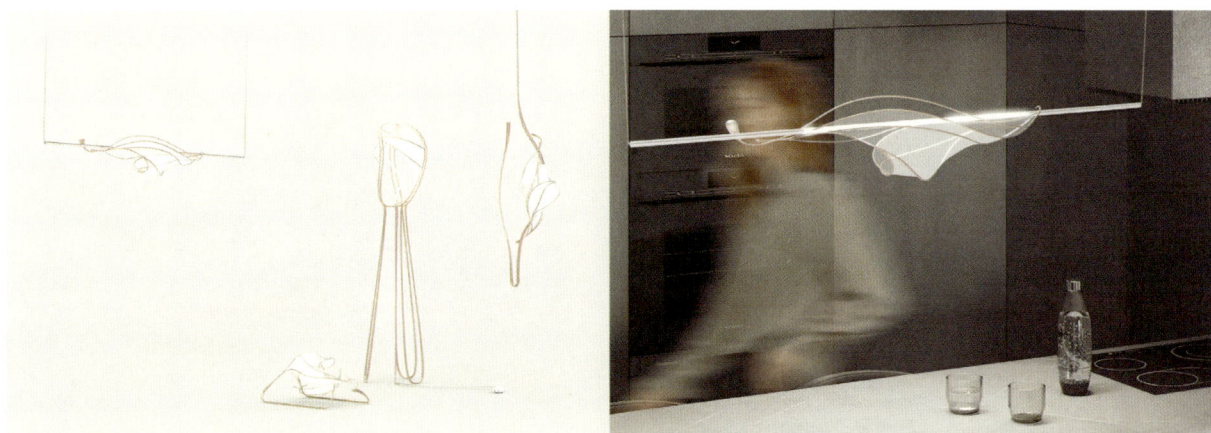

↘ 2.2.2 课题：如何优雅地运动

2.2.2 TASK: HOW TO EXERCISE GRACEFULLY

课题说明：

　　全球的健康促进和体育活动在实践中均受到极大的冲击。因此，思考如何保证民众的生活质量，并维持与确保永续发展目标的达成，成为全球健康促进的新议题。一些国家的公共政策发生了改变，如英国把人行道及脚踏车道拓宽，以增加民众运动的机会，避免民众自驾与公交车或捷运相拥挤。2021年中国台湾的一项运动现况调查发现，一方面，民众更关注身体健康，规律运动人口比率创新高；另一方面，民众改变运动模式，在家运动的比率从去年的9%增至15.3%，许多企业也开始推广如何将运动融入生活而养成"运动好习惯"。除了勤洗手、保持社交距离外，通过维持运动习惯，提高身体免疫力，才是每个人都必须做好的功课。

指导老师：

　　黄文宗，中原大学商业设计系副教授，设计学院博士生导师
　　范宝莲，领袖人整合行销有限公司创办人

Task Description:

　　Global health promotion and physical activities have been greatly impacted. Therefore, thinking about how to maintain people's quality of life, and ensure the achievement of sustainable development goals has become a new topic of global health promotion. The public policies of some countries also changed. For example, the United Kingdom widened the sidewalks and footpaths to increase the opportunities for people to exercise more, and they did not need to squeeze buses or MRTS. In 2021, a survey on the current situation of sports in Taiwan found that people pay more attention to their health, and the proportion of people exercising regularly has hit a new high; people have changed their exercise patterns, and the proportion of exercising at home has increased from 9% last year. By 15.3%, many companies have also begun to promote how to incorporate exercise into their lives to develop "good exercise habits." In addition to washing hands regularly, keeping social distance, maintaining exercise habits, and building body immunity are the lessons everyone must follow.

Tutor:

　　Wenzong Huang, Associate Professor, Department of Business Design, Chung Yuan Christian University, Doctoral Supervisor of the School of Design

　　Baolian Fan, Founder of Leaders Integrated Marketing Co., Ltd.

MOHO——基于柔性交互的居家智能运动辅助产品设计

MOHO—Intelligent Product Design Based on Flexible Interaction For Home Motion Assistance

作者：韩晴、刘建月、龙腾明、佘鼎妍、王贝贝、张嘉珩

By Qing Han, Jianyue Liu, Tengming Long, Dingyan She, Beibei Wang, Jiaheng Zhang

MOHO，Move At Home，是优雅运动的新定义。MOHO 从日常居家中最常见的瑜伽垫、护膝、护腰等运动辅助设备入手，结合优雅的智能技术，创造智慧化的运动体验。科技不应该是冰冷的，所以我们采用优雅自然的柔性交互方式，选用布料等柔软织物材料，让用户在运动过程中获得更加亲近自然的体验和优雅的生活状态。人们的居家健身态度正朝着更加积极的方向改变，我们将为居家健身用户提供个性化的居家运动辅助和专业化的运动过程指导，并通过趣味化和激励化的方式为居家健身用户提供即时的运动反馈。

MOHO, Move at Home, is a new definition of elegant exercise. MOHO starts with the most common exercise AIDS in the daily home, such as yoga MATS, knee pads, and wrist protection. It combines elegant intelligent technology to create an intelligent exercise experience. Science and technology should not be cold and raw, so we adopt the elegant and natural flexible interaction method and choose soft fabric products and materials so that users can have a closer-to-nature experience in the process of sports, and an elegant living state. After the era of the outbreak of home fitness is changing in the direction of a more positive attitude, we provide personalized home exercise assistance and professional exercise process guidance for home fitness users and provide instant exercise feedback to home fitness users in a fun and motivating way.

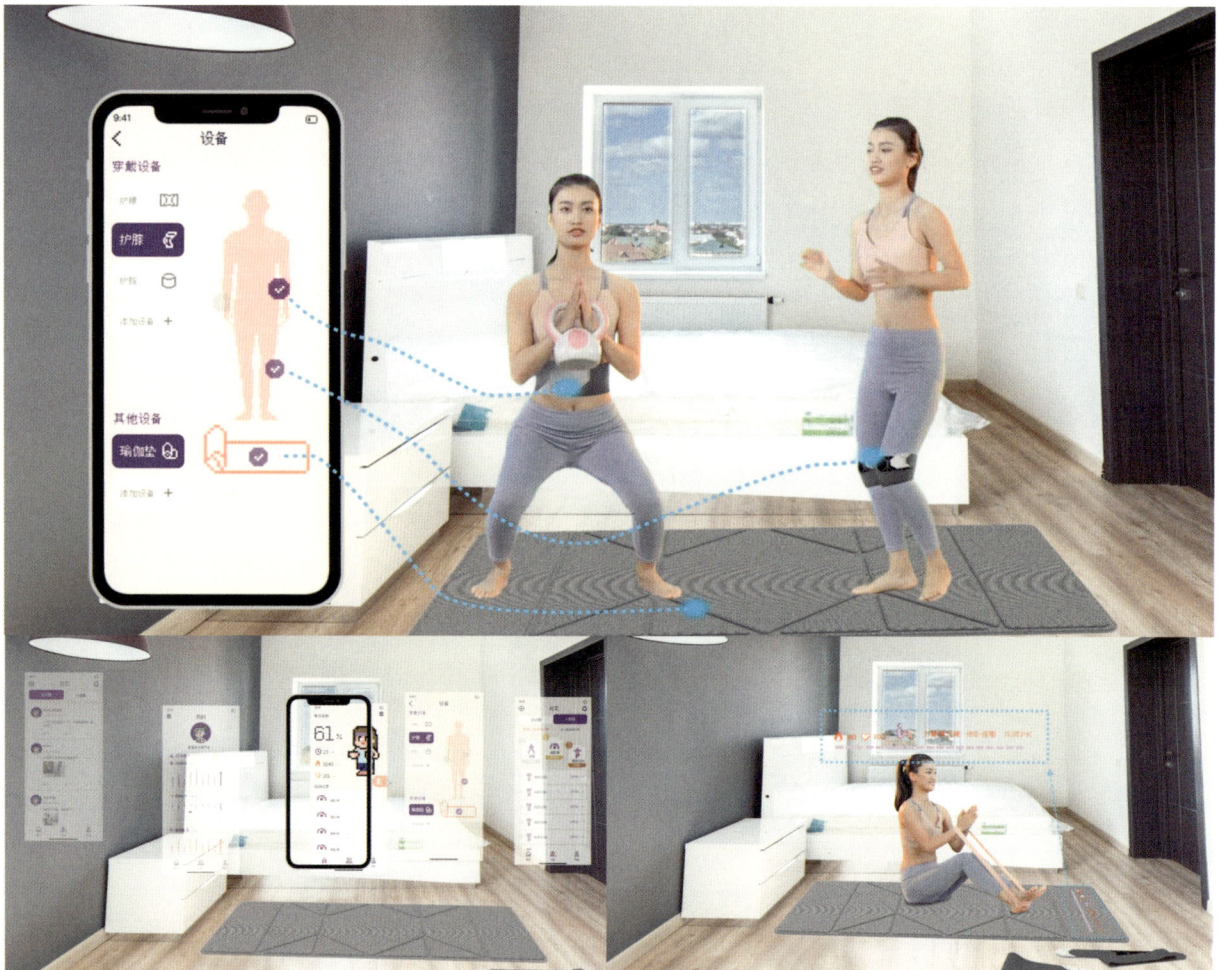

↘ **2.2.3　课题：我们开始吧——优雅的行为**

2.2.3　TASK: LET'S DO IT—GRACEFUL BEHAVIOR

课题说明：

　　行为贯穿于我们生活的衣食住行，不同的行为会产生不同的影响。比如高盐的饮食行为可能会导致慢性疾病，甚至影响生命长度；乱丢垃圾的行为会影响生态环境，污染土壤及水资源；有趣的空间设计，可以让行人加快步行速度，减少交通拥堵。通过设计改变用户的行为，对于设计师来说是一个极大的挑战，不仅要求设计师有强大的设计能力，而且需要设计师能够洞察用户动机，用设计的方式来触发用户行为的改变。本课题希望设计师可以通过简单而有力量的设计方案引导或改变用户的行为，使用户产生"优雅的行为"，通过行为的设计来解决社会性难题。

Task Description:

　　Behavior runs through our daily lives. Different behaviors will have different effects. For example, a high salt diet may lead to chronic diseases and even affect the length of life. Littering will impact the ecological environment and pollute oil and water resources. The attractive spatial design will speed up the walking speed of pedestrians and reduce traffic congestion. It is a great challenge for designers to change behavior through design, which requires designers to have strong design ability and insight into user's motivation and trigger the change of user behavior by design. Through this task, we hope that the designers can guide or change the user's behavior through simple and robust design schemes so that the user can have "graceful behavior" and solve a social problem through behavior design.

指导老师：

　　马官正，南京信息工程大学数字媒体艺术系讲师

　　高原，南京信息工程大学设计艺术系副教授

　　梅凌婕，美好心理（北京）科技有限公司CEO、创始人

Tutor:

　　Guanzheng Ma, Lecturer of the Department of Digital Media Art, Nanjing University of Information Science and Technology

　　Gao Yuan, Associate Professor, Department of Design and Art, Nanjing University of Information Science and Technology

　　Lingjie Mei, CEO and Founder of Good Psychology

2.2.3.1　Let's 狗

2.2.3.1　Let's GO

作者：孙靖涵、蔡子绮、李俊男、罗瑀雯、王筠妤、杨启航

By Jinghan Sun, Ziqi Cai, Junnan Li, Yuwen Luo, Yunyu Wang, Qihang Yang

　　宠物已经成为很多人生活中不可分割的一部分，2021年中国宠物狗数量已经达到了5800万只，但是现在越来越多的宠物狗生活环境已经大打折扣，高压生活下的宠物主人陪伴宠物的时间逐渐减少，宠物独自在家也会产生心理疾病；社会对于宠物的接受度参差不齐，宠物狗有时候也会成为受歧视的群体。我们根据现状和痛点，希望以北京的口袋公园为切入点，打造宠物友好型空间场地。

　　Pets have become an integral part of many people's lives. In 2021, the number of pet dog grooming in China reached 58 million. However, the living environment of more and more pet dogs has been greatly compromised, and pet owners need more time to accompany their pets under high pressure. In addition, the social acceptance of pets varies widely, and pet dogs sometimes become a stigmatized group in society. According to the current situation and pain points, we plan to take the pocket park in Beijing as the starting point to create a pet-friendly space specially prepared for pets.

2.2.3.2 "树懒进食计划"——智力缺陷儿童的饮食认知

作者：陈宝成、胡徐妍、莫洁琼、孙飞雪、孙佳钰、杨芷瑄

　　我们把智力缺陷儿童比喻成"树懒宝宝"，希望针对树懒宝宝的饮食认知问题制订"树懒进食计划"，增强树懒宝宝对于食物的认知。其中包括实体的食材餐盘、食物模拟组件、大口吃饭游戏机，运用艺术疗法和四个趣味小游戏，对其五感进行训练，建立与食物的对应关系，色彩、材质与外形的呈现满足孩子对于具象物品认知的需求。同时搭配线上平台，通过线上应用和线下活动，搭建物品交换、家长交流、儿童交友的可持续交流平台。

2.2.3.2 "Sloth Eating Program"—Dietary Cognition of Children with Intellectual Disabilities

By Baocheng Chen, Xuyan Hu, Jieqiong Mo, Feixue Sun, Jiayu Sun, Zhixuan Yang

　　We compare children with intellectual disabilities to "sloth babies" and hope to formulate a "sloth eating plan" for the sloth baby's diet cognition to enhance the baby's cognition of food. It includes the physical food plate, food simulation component, big mouth eating game machine, art therapy, four fun games, training the five senses, and establishing the corresponding relationship with food, color, material, and appearance to meet the children's cognitive needs for concrete objects. At the same time, through online applications and offline activities, a sustainable communication platform for Commodity Exchange, parent communication, and children making friends is constructed.

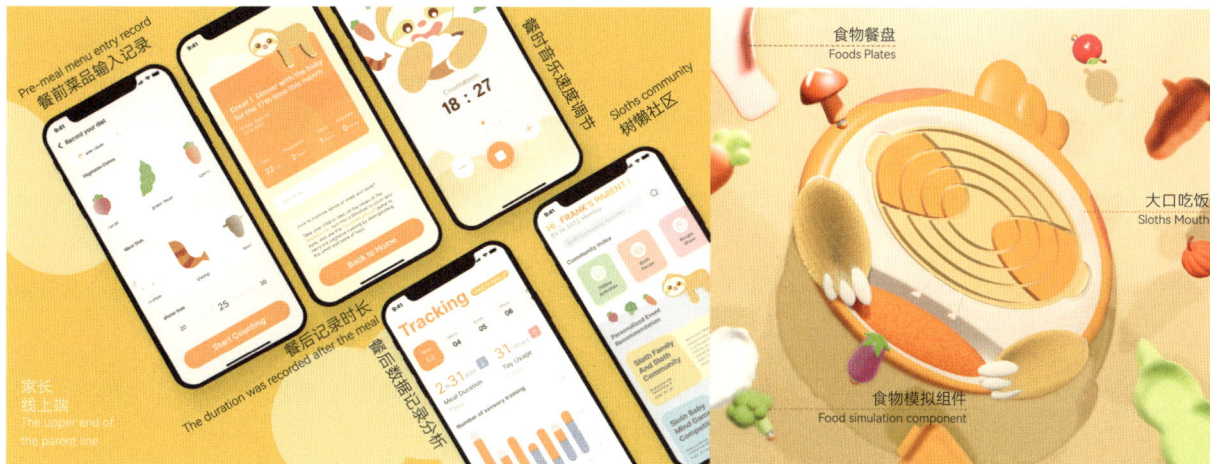

↘ 2.2.4　课题：老龄化社会背景下的公共产品设计

2.2.4　TASK: PUBLIC PRODUCT DESIGN IN THE CONTEXT OF AN AGING SOCIETY

课题说明：

　　当前，中国的社会老龄化程度已经以各项严峻的数据体现出来，这种情势的具体影响已经体现为结构性问题。我们当代的设计教育基本根植于消费设计，它在面对结构性社会需求时已显不足。阿瑟·克莱曼（Arthur Kleinman）团队（哈佛大学）在社会科技方法中提出社会运行和文化环境两个主要方面的综合选择和综合发展。在多学科团队的合作研究中找到问题的根源，并制定基于个人、家庭和社会的解决方案来面对这一系列问题。我们基于此提出"Social Product Design"的方法，并在江苏省产业技术研究院的适老科技创新中心的一系列项目中实践运用，本次课程也是该课题持续研究实践的一部分。

指导老师：

　　柏雷，南京艺术学院副教授，江苏省产业技术研究院集萃适老设计中心主任

　　江加贝，南京艺术学院工业设计学院产品设计系副教授

Task Description:

The current aging of Chinese society is reflected in grim statistics. The concrete impact of this situation is already reflected in structural issues. Our contemporary design education, primarily rooted in consumer design, needs to be improved in the face of structural social needs. Arthur Kleinman's team (Harvard University) proposes an integrated choice and development of the two main aspects of social functioning and cultural environment in a social technology approach. In a collaborative multidisciplinary team approach, we identify the root causes of problems and develop individual, family, and social solutions to various issues. This course is part of the ongoing research and practice of the "Social Product Design" approach, which is being applied in several tasks of the Center for Age-Friendly Technology and Innovation of the Jiangsu Institute of Industrial Technology.

Tutor:

Lei Bai, Associate Professor, Nanjing University of the Arts; Director of the Design Center for the Elderly, Jiangsu Industrial Technology Research Institute

Jiabei Jiang, Associate Professor, Department of Product Design, School of Industrial Design, Nanjing University of the Arts

老龄化社会背景下的公共产品设计

Public Product Design Under the Background of Aging Society

作者：杨若望、李芷芸、李红丽、李可欣、郝佳淑

By Ruowang Yang, Zhiyun Li, Hongli Li, Kexin Li, Jiashu Hao

养老问题是目前政府和社会亟待解决的热点问题，而洗浴问题更是保障老人基本生活质量的一大重要因素。我们发现，很多半失能老人、失能老人普遍存在洗澡难、养护难的问题，会在生理、心理等方面，对老人的养老生活质量造成负面影响。我们希望利用社会性产品以及服务设计来帮助老人解决这一问题。通过社会性产品与服务设计，我们开发了一套符合老人助浴流程的产品，并建立了一条新的、家与社区的有机连接。从社交、养生、娱乐等方面为老人洗浴问题提供解决方案，也是为解决老龄化社会问题贡献力量，衷心希望每一个老人都能安享晚年。

The problem of providing for the elderly is a hot issue that the government and society urgently need to solve, and the problem of bathing is a major factor in ensuring the basic quality of life of the elderly. We found that many semi-disabled elderly and disabled elderly generally have difficulties in bathing and maintenance, which will have a bad impact on the elderly's quality of life in terms of physiology and psychology. We hope to use social products and service design to help the elderly solve this problem. Through social product and service design, we have designed products that conform to the elderly's bath aid process and established a new organic connection between home and community. Contributing to the problem of bathing for the older adults in terms of socialising, health and recreation is also contributing to the problem of an aging society, and it is sincerely hoped that every older people can enjoy a peaceful life in their old age.

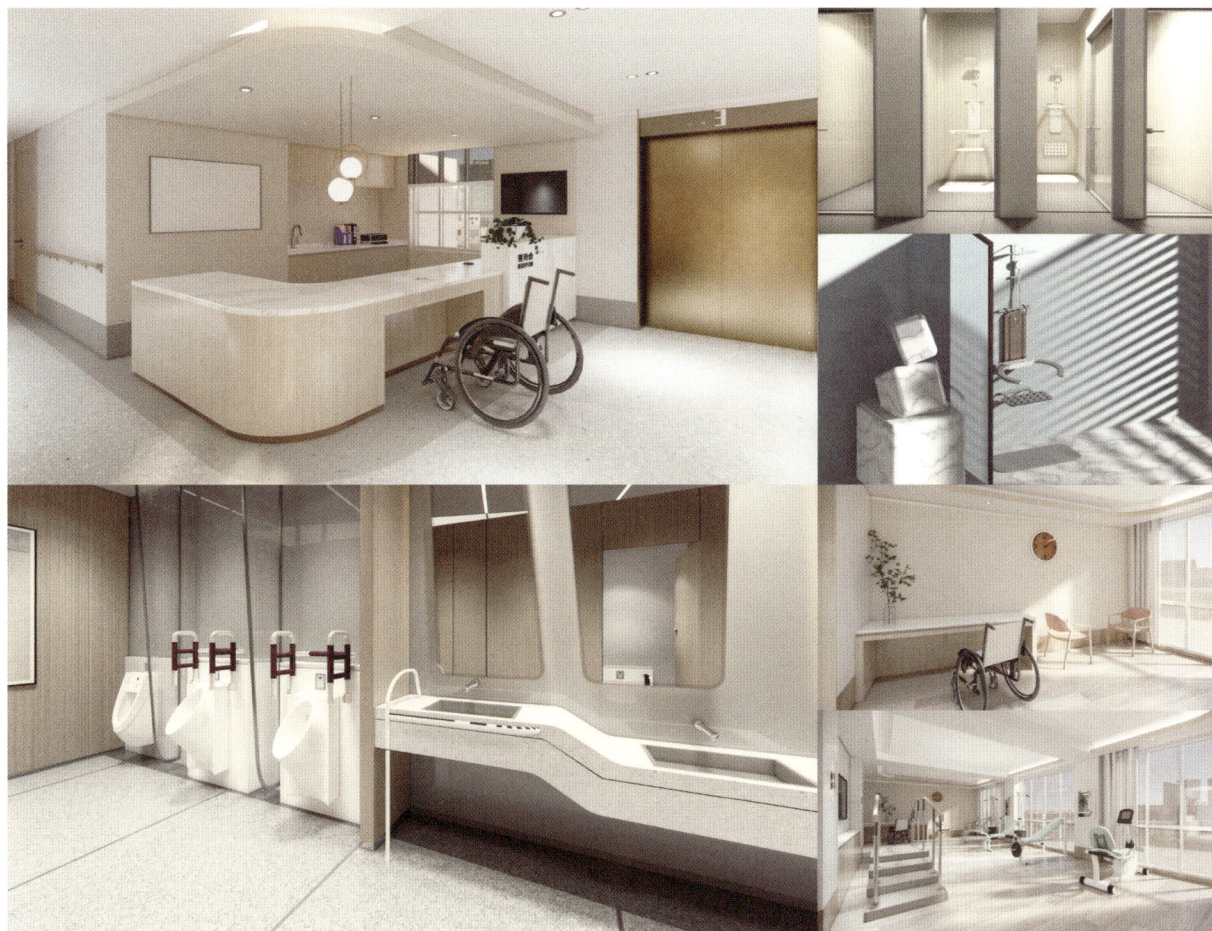

↘ **2.2.5 课题：老龄社区服务系统设计**

2.2.5 TASK: AGING COMMUNITY SERVICE SYSTEM DESIGN

课题说明：

中国老龄化趋势和老龄化所带来的一系列问题在近年来受到广泛关注。如何真正了解老年人生活中遇到的问题，如何使设计真正满足老年人的需求，是我们面临的挑战。协同设计是设计师、用户与其他利益相关者共同参与设计实践的设计方法，通过集体的思考和行动，解决如今面临的挑战。协同设计鼓励我们利用生成性工具包等方法与最终的消费者和用户共同行动，从而达到价值共创。

指导老师：

汪晓春，北京邮电大学副教授

Task Description:

The trend of aging in China and the range of issues that aging brings has received much attention in recent years. The challenge is how to truly understand the problems encountered in the lives of older people and how to make designs that truly meet their needs. Codesign is a design approach where designers, users, and other stakeholders participate in design practice together, solving today's challenges through collective thinking and action. Codesign encourages us to act with end consumers and users using generative toolkits, leading to value creation.

Tutor:

Xiaochun Wang, Associate Professor, Beijing University of Posts and Telecommunications

2.2.5.1　手牵守——老人接娃"共享幸福"服务系统设计

2.2.5.1　Love In Hand—Upgrade the Happiness Of Elders Picking Up Children Service System

作者：沈柯蓉、梁哲龙、潘宇杰、张宇、李雅

By Kerong Shen, Zhelong Liang, Yujie Pan, Yu Zhang, Ya Li

　　"手牵守"是以"共享幸福"为核心，以便捷和舒适为目标，针对需要接送孙辈的老人群体，进行的老人接娃服务系统设计，重新定义祖孙相互陪伴的这段特殊路途，为祖孙提供一个安全的接送环境，为老人设置舒适的等待区域和购物区，将甜蜜的负担转变为享受幸福的过程，同时设置了亲子活动区域，让共享幸福的时光更加值得记忆。

　　"Love in Hand" is a service system designed for older people who need to pick up and drop off their grandchildren, with "shared happiness" as the core and convenience and comfort as the goal. It creates safe pick-up and drop-off environment, a comfortable waiting area, and a shopping area for older people, transforming the sweet burden into a process of enjoying happiness while setting up a parent-child activity area to make the time of sharing joy more memorable.

2.2.5.2　智汇·老龄社区防诈生活系统设计

作者：吴凯丰、方伟彤、周珺谕、蔡钰珊、周心惟

智汇·老龄社区防诈生活系统致力于重新定义老人防诈认知，开启智能防诈新模式。解决老人与子女的共同忧虑，让生活更加优雅从容。该系统拥有物理触点、媒介触点以及人际触点，分别对应的设计点为数字宣传显示屏、防诈社区情景体验装置以及结对家庭。居民们能够通过互动得到反诈故事卡，起到进一步反诈宣传作用。另外，结对家庭能够带动与温暖独居老人，不仅能起到防诈作用，还可以进一步减少他们的生活孤独感。

2.2.5.2　Design Of Anti-Fraud Living System In Elderly Community

By Kaifeng Wu, Weitong Fang, Junyu Zhou, Yushan Cai, Xinwei Zhou

The design is committed to a new definition of elderly anti-fraud cognition, opening a new model of intelligent anti-fraud. To solve the typical worries of the elderly and children, make life more elegant and effortless. The system has physical contact touchpoints, media contact touchpoints, and interpersonal touchpoints. The corresponding design points are a digital advertising display screen, an anti-fraud community scene experience device, and paired families. Residents can get anti-fraud story cards through interaction to further promote anti-fraud. In addition, paired families can stimulate and warm the elderly living alone, preventing fraud and further reducing their loneliness in life.

服务流程设计
Service Process Design

案例实战情景模拟
Case real combat scenario simulation

宣传与公示
Publicity and publicity

价值力对比／竞争
Value force comparison/competition

成员晋升
Member promotion

社区公共体验装置
Community public experience device

智能宣传栏
Intelligent bulletin board

吸纳更多组员
Recruit more team members

老人产生疑问
The elders have questions

记录最新骗局
Record the latest scam

AI骗局库
AI Scam Library

子女分享最新案例
Children share the latest case

2.2.6 课题：如何优雅地时尚——西南非遗的数字化时尚设计

2.2.6 TASK: HOW TO MAINTAIN FASHION GRACEFULLY—DIGITAL FASHION DESIGN OF SOUTHWEST INTANGIBLE CULTURAL HERITAGE

课题说明：

　　数字时尚是近年来最新、最热的主题之一，并不断呈现出大众化、国际化和年轻化趋势。中国西南地区蕴藏着丰富的非物质文化遗产。在课题中，我们将尝试通过设计创新方法，突破空间、材质、创作维度、交流与展示等方面的限制，为其创造性转化与创新性发展提供思路。导师组目前已联合西南非遗传承人、地方政府、文旅企业等完成多项基于民族民间传统文化及工艺的创新展演活动，并输出了大量设计创新作品，有较好的研究基础与数据，希望能够与同学们一起，共同探索西南非遗数字化时尚创新的新方法、新宇宙。

指导老师：

　　蔡端懿，四川大学艺术学院讲师
　　杨璐铭，四川大学服装与服饰设计系主任
　　许亮，副教授，四川大学艺术学院设计与媒体艺术系主任
　　熊彦，四川吾思吾行文化产业有限公司创始人

Task Description:

　　Digital fashion is one of the newest and hottest themes in recent years and continues to show popular, international, and youthful trends. China's southwestern region is rich in intangible cultural heritage. In the task, we will provide ideas for its creative transformation and innovative development through design innovation methods to break through space, material, creation dimension, communication, and display limitations. The tutor group has completed several innovative exhibition activities based on ethnic folk traditional culture and craftsmanship in collaboration with Southwest African hereditary inheritors, local government, and cultural tourism enterprises. It has outputted many design innovation works, with a good research base and data, hoping to explore new methods and universes of digital fashion innovation of South West African heritage together with the students.

Tutor:

　　Duanyi Cai, Lecturer at the College of Arts, Sichuan University

　　Luming Yang, Dean of the Department of Fashion and Apparel Design, Sichuan University

　　Liang Xu, Associate Professor, Dean of the Department of Design and Media Art, College of Arts, Sichuan University

　　Yan Xiong, Founder of Sichuan Wusi Wuxing Cultural Industry Co., Ltd.

2.2.6.1 寻遗

2.2.6.1 Searching for Intangible Cultural Heritage

作者：邵晓倩、李心竹、周佳俊、闫欣、赵倩、张泽

By Xiaoqian Shao, Xinzhu Li, Jiajun Zhou, Xin Yan, Qian Zhao, Ze Zhang

运用数字化技术构建线上非遗世界，以中国西南地区代表性非遗为依托，打造一个虚拟的非遗社区平台。在虚拟的非遗社区，拥有以不同非遗元素为主题的村落。用户通过APP的使用，可以了解、观看、学习非遗相关知识，让西南地区非遗得到更广的传播，并从侧面进行传承。通过数字化技术搭建虚拟社区平台，赋予西南非遗新的生命力，为西南非遗的创新型发展提供新的思路。

Build an online intangible cultural heritage world with digital technology and build a virtual intangible cultural heritage community platform based on the representative intangible cultural heritage in southwest China. In the virtual intangible cultural heritage community, there are villages with different intangible cultural heritage elements as the theme. Using the app, users can understand, watch, and learn about intangible cultural heritage so that intangible cultural heritage in southwest China can be more widely spread and passed on laterally. Through digital technology, a virtual community platform is built to give new vitality to the Southwest intangible cultural heritage and provide new ideas for the innovative development of Southwest intangible cultural heritage.

公共平台

个人平台

非遗世界绘画设计及建模
Painting Design and Modeling of Intangible Cultural Heritage World

2.2.6.2 声声不息

2.2.6.2 Everlasting Echo

作者：章悦、张雪松、胡蒙、陈鑫、刘子珊、温玉婷

By Yue Zhang, Xuesong Zhang, Meng Hu, Xin Chen, Zishan Liu, Yuting Wen

我们将贵州非遗四十八寨歌节、玉屏箫笛、铜鼓和芦笙的声音进行视觉化呈现，运动的图案参考了贵州的人文地理，使用水元素作为视觉展示，水纹随着声音而律动。用创新和跨界的形式传播非遗背后的社会价值，希望人们享受非遗乐器的声音，看到非遗的有趣，听到非遗的多样性。在非遗的声音中，感受不息的力量，让我们的生活变得更优雅。

We visually present the sounds of the Song Festival of the 48zhai Village in Guizhou Intangible Cultural Heritage, Yuping Flute, Bronze Drum, and Lu Sheng. The movement patterns refer to the human geography of Guizhou, and the element of water is used as the visual display, and the water lines are rhythmic with the sounds. The social value behind the intangible cultural heritage should be spread in innovative and cross-border forms, hoping that people can enjoy the sound of intangible cultural heritage instruments, see the interesting, and hear the diversity of intangible cultural heritage in the anxious post-pandemic era. Feel the ceaseless power in the sound of intangible cultural heritage and let our lives become more elegant.

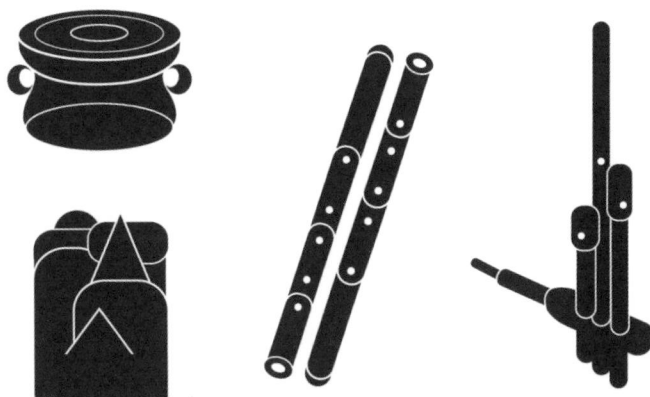

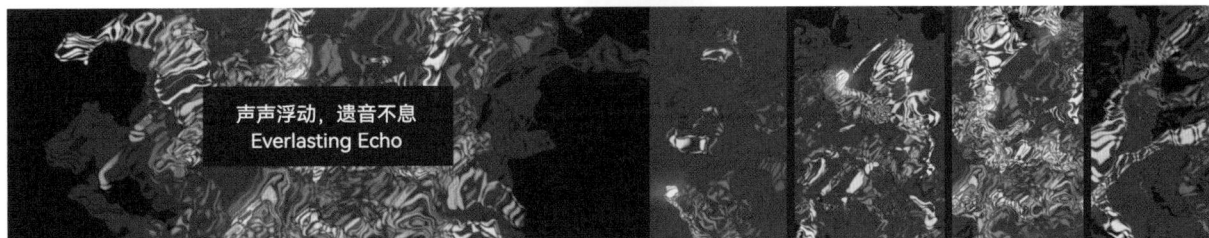

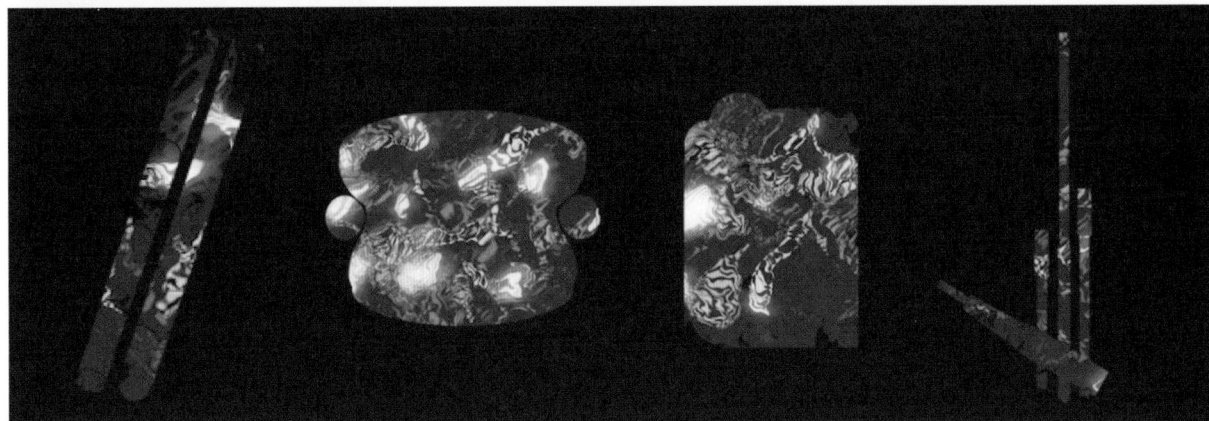

↘ **2.2.7 课题：乡村持续力 —— 艺术设计赋能乡村振兴**

2.2.7 TASK: RURAL SUSTAINABILITY—ART & DESIGN EMPOWERS RURAL VITALIZATION

课题说明：

目前超过一半的地球居民居住在城市，乡村空心化造成城乡分割严重。2021年4月，《中华人民共和国乡村振兴促进法》正式审议通过，乡村振兴包括产业振兴、人才振兴、文化振兴、生态振兴和组织振兴。课题将深入探索乡村，与工作在乡村振兴一线的工作人员和村民们进行共创。基于乡村发展现状与需求，聚焦数字化乡村、乡村文旅IP、乡村可持续环境营造、乡村养老等多个领域，产出能发展于乡村、振兴于乡村、落地于乡村的服务策略和不同形式的文化触点。通过艺术、服务设计、产品设计、视觉设计等帮助乡村进行产业振兴、文化振兴、生态振兴。

Task Description:

Nowadays, more than half of the earth's inhabitants currently live in cities, and the hollowing out of the countryside has resulted in a solid urban-rural divide. In April 2021, the country introduced the Law of the People's Republic of China on the Promotion of Rural vitalization, which includes industrial vitalization, talent vitalization, cultural vitalization, ecological vitalization, and organizational vitalization. The task will explore the countryside in depth and realize co-creation with the staff and villagers working on the front line of rural vitalization. Based on the combination of the current situation and needs of rural development, the project focuses on various fields such as the digital village, rural cultural tourism IP, rural sustainable environment creation, rural retirement, etc. It produces service strategies and different forms of cultural contacts that can be developed in the village, revitalized, and landed in the village. Through art, service design, product design, and visual design, we help villages to revitalize their industries, culture, and ecology.

指导老师：

赵璐，教授，博士生导师，鲁迅美术学院副院长

张超，鲁迅美术学院中英数字媒体艺术学院讲师

尹香华，鲁迅美术学院中英数字媒体艺术学院讲师

毕卓异，鲁迅美术学院中英数字媒体艺术学院讲师

Tutor:

Lu Zhao, Professor, Doctoral Supervisor, Vice President of LuXun Academy of Fine Arts

Chao Zhang, Lecturer of Sino-British Digital Media Art School, LuXun Academy of Fine Arts

Xianghua Yin, Lecturer of Sino-British Digital Media Art School, LuXun Academy of Fine Arts

Zhuoyi Bi, Lecturer of Sino-British Digital Media Art School, LuXun Academy of Fine Arts

2.2.7.1　收获鱼趣，做一回大海的孩子

作者：刘东辉、宫菲繁、顾函珏、张子豪、郑斯茵、江欣忆

　　结合广鹿岛赶海体验，我们推出吸引年轻群体的"鱼趣"，这款伴随全流程的应用程序，具备创新性、先锋性和较强的趣味体验性，紧扣广鹿的历史渊源而设计。在曾经的广鹿岛有许多结对的鹿群，因此，我们推出了属于广鹿岛的"艺术村长"——"岛鹿鹿"。在用户游览广鹿岛的同时，"岛鹿鹿"以不同的形象和身份时刻陪伴在用户的身边，在赶海体验过程中一起学习赶海知识和小技巧。并通过AR技术增强赶海游戏体验，通过实景扫描召集七位系列虚拟赶海村长将伴随整个赶海过程。

2.2.7.1　Harvest Fishing Fun and Be a Child of The Sea

By Donghui Liu, Feifan Gong, Hanyu Gu, Zihao Zhang, Siyin Zheng, Xinyi Jiang

　　In combination with the experience of sailing to the sea on Guanglu Island, we launched a series of innovative, pioneering, and fun-enhancing experiences to attract young groups. The whole process is accompanied by the APP "Fishing Fun for According to the historical reasons." Guanglu Island, there were many pairs of deer there. In combination with the above elements, we launched the "art village head" for Guanglu Island—"Island Deer". While visiting Guanglu Island, "Island Deer" will always accompany you with different images and identities, and learn the knowledge and tips of sea-going together during the experience. Through AR augmented reality sea-going game experience, a series of seven virtual sea-going village heads were convened through live scene scanning to accompany the whole sea-going process.

Fishing Fun/ 鱼趣 App

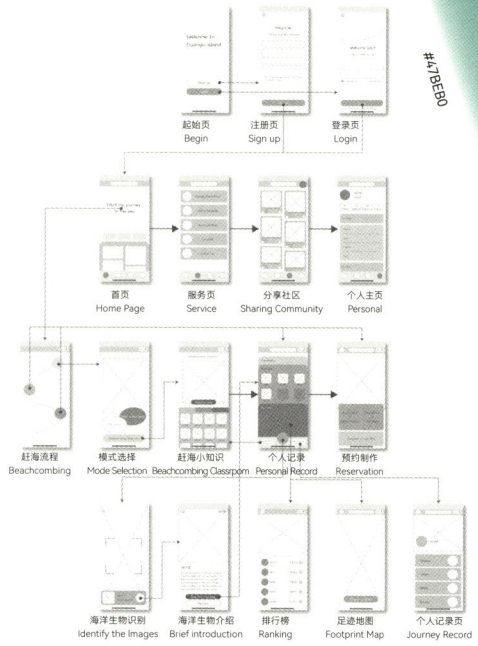

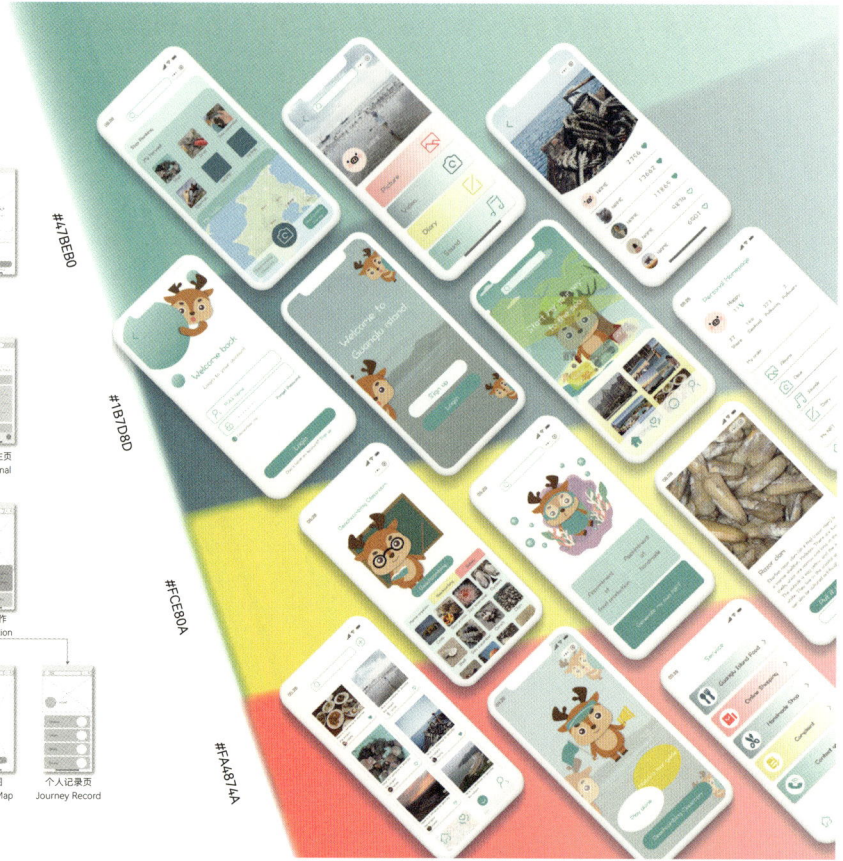

起始页 Begin
注册页 Sign up
登录页 Login

首页 Home Page
服务页 Service
分享社区 Sharing Community
个人主页 Personal

赶海流程 Beachcombing
模式选择 Mode Selection
赶海小知识 Beachcombing Classroom
个人记录 Personal Record
预约制作 Reservation

海洋生物识别 Identify the Images
海洋生物介绍 Brief introduction
排行榜 Ranking
足迹地图 Footprint Map
个人记录页 Journey Record

#47BEB0
#1B7D8D
#FCE80A
#FA4874A

2.2.7.2 广鹿岛乡村振兴——来吧，做一天岛主大人吧!

作者：王溢婧、苏敬文、李彤、范思宇、徐帅、李俐婷

"鹿岛游"旨在为青年旅游群体提供现捞现吃、随地可住、定制路线的沉浸式广鹿岛特色旅行体验，让他们在旅游时能实现"躺平"，解决年轻群体前往广鹿岛旅游体验感差、同质化严重、旺季住宿预定困难及价格高的问题，从而开拓广鹿岛旅游的新奇体验。以吃住游一体的"躺平"APP、便携的工具包和身份标识手环为载体，为"广鹿岛主大人"构建了轻松趣味的、原汁原味的广鹿岛游玩服务体系。

2.2.7.2 Rural Vitalization on Guanglu Island—Come and Be an Islander for a Day

By Yijing Wang, Jingwen Su, Tong Li, Siyu Fan, Shuai Xu, Liting Li

"Guanglu Island Tour"aims to provide young travellers with an immersive travel experience on Guanglu Island with ready-to-eat food, accommodation anywhere, and customised routes, so that they can. "lie down" when travelling, and to solve the problems of poor experience, selias homogeneity, difficulty in booking accommodation in peakseasons and high prices for the young travelling to Guanglu Island, so as to open up novel experiences for them. This task uses the Lie Flat APP, portable tool kit, and identity bracelet as the carrier to build a relaxed, interesting, and original Guanglu Island play service system for "Guanglu Island Master".

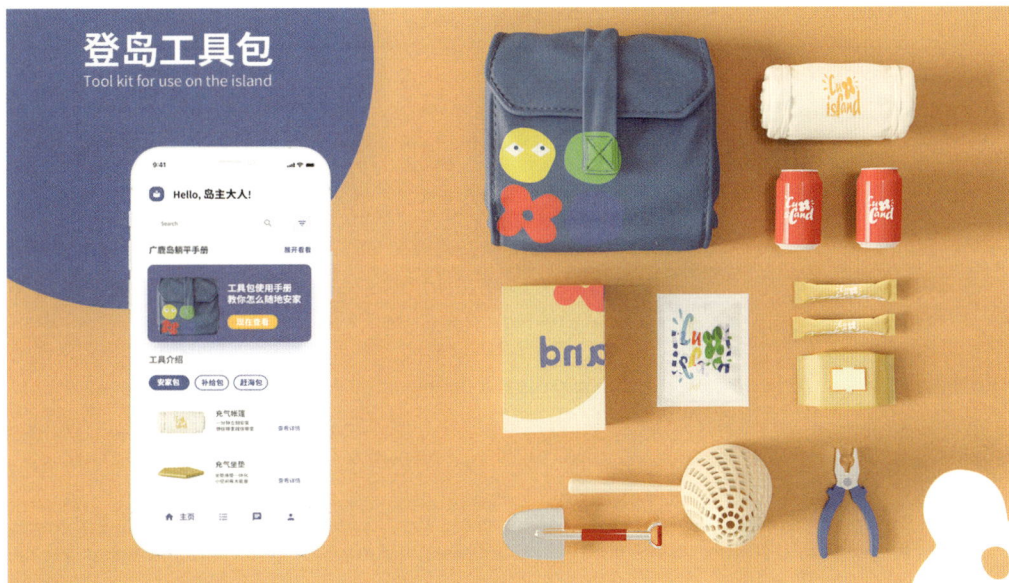

↘ **2.2.8　课题：虚拟时尚拟态设计**

2.2.8　TASK: VIRTUAL FASHION MIMETIC DESIGN

课题说明：

伴随着新一代数字技术的加速演进，全球数字资产数量呈现出井喷式暴增趋势，虚拟技术已在我们周遭驱动了巨大的产业变革，而作为青年一代的你，对虚拟时尚的未来有怎样的创想和思考呢？本课题基于"拟态化设计"思维和方法，将现实世界作为拟态研究对象，探讨虚拟世界中"人、物、空间构成关系""虚拟和现实孪生关系转化"等问题，引导课题成员建立多维数字观，理解虚拟设计作为独立语言所特有的差异性，思考新技术产生背景下的技术伦理，感受虚拟与现实的边界以及其作为媒介如何改变人类文化的认知方式。

指导老师：

魏勤文，北京服装学院服饰艺术与工程学院虚拟时尚设计方向行政主任

周小凡，北京服装学院服饰艺术与工程学院产品设计系教师

王涛，北京服装学院服饰艺术与工程学院虚拟时尚教学部专业教师

贺爽，清华大学设计学在读博士研究生

刘源，北京服装学院虚拟时尚系教师

崔艺铭，北京服装学院服饰艺术与工程学院产品设计系助理教授

Task Description:

With the accelerated evolution of the new generation of digital technology, the number of digital assets around the world has shown an explosive increase, and virtual technology has driven massive industrial changes around us. As a young generation, what are your ideas and thoughts on the future of virtual fashion? This task is based on "mimetic design" thinking and methodology, taking the real world as the object of mimetic research, exploring "the relationship between human, object and space composition in the virtual world" "the twin relationship between virtual and real transformation". The task members are guided to establish a multidimensional digital view, understand the differences of virtual design as an independent language, consider the ethics of technology in the context of new technologies, and feel the boundary between virtual and real and how it changes the way human culture is perceived as a medium.

Tutor:

Qinwen Wei, Executive Dean of Virtual Fashion Design Direction, School of Fashion Art and Engineering, Beijing Institute of Fashion Technology

Xiaofan Zhou, Teacher of Product Design, School of Fashion Art and Engineering, Beijing Institute of Fashion Technology

Tao Wang, Professional teacher, Virtual Fashion Teaching Department, School of Fashion Art and Engineering, Beijing Institute of Fashion Technology

Shuang He, A Ph.D. candidate in Design, Tsinghua University

Yuan Liu, Teacher of Virtual Fashion, Beijing Institute of Fashion Technology

Yiming Cui, Assistant Professor of Product Design, School of Fashion Art and Engineering, Beijing Institute of Fashion Technology

2.2.8.1　虚拟时尚设计下的数据拟态首饰表达

2.2.8.1　Data Mimetic Jewelry Expression Under Virtual Fashion Design

作者：陈缘圆、陈煊轩、宋德奥、谢嘉颖、崔颖、王宁

By Yuanyuan Chen, Xuanxuan Chen, De'ao Song, Jiaying Xie, Ying Cui, Ning Wang

只要拥有足够多的数据记录，就可以拼凑出一个人的"数据轮廓"，并预测出这个人的教育经历、消费理念、兴趣爱好……这就是如今网络世界的奇幻现实。我们以此为灵感，将数据拟态与虚拟时尚结合，设计了"10101010"系列虚拟首饰，表现数据如空气般充斥在我们生活的世界的现状，让大家去体验、拥抱、反思或是批判这一现象。

With enough data records, it is possible to piece together a "data profile" of a person and predict their education, consumption, interests, and hobbies... This is the fantasy reality of today's online world. Inspired by this, we combined data mimicry with virtual fashion and designed the "10101010" series of virtual jewelry. It's an expression of the air of data in our world. Let everyone experience, embrace, reflect, or criticize the phenomenon.

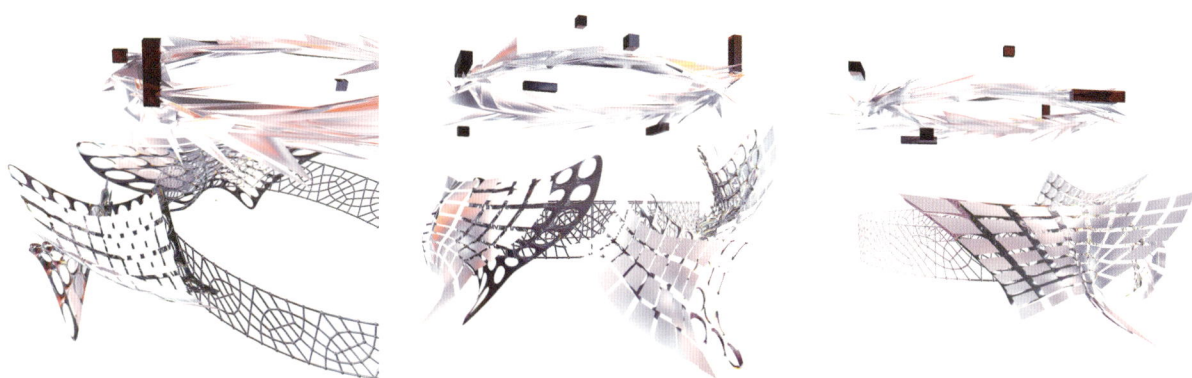

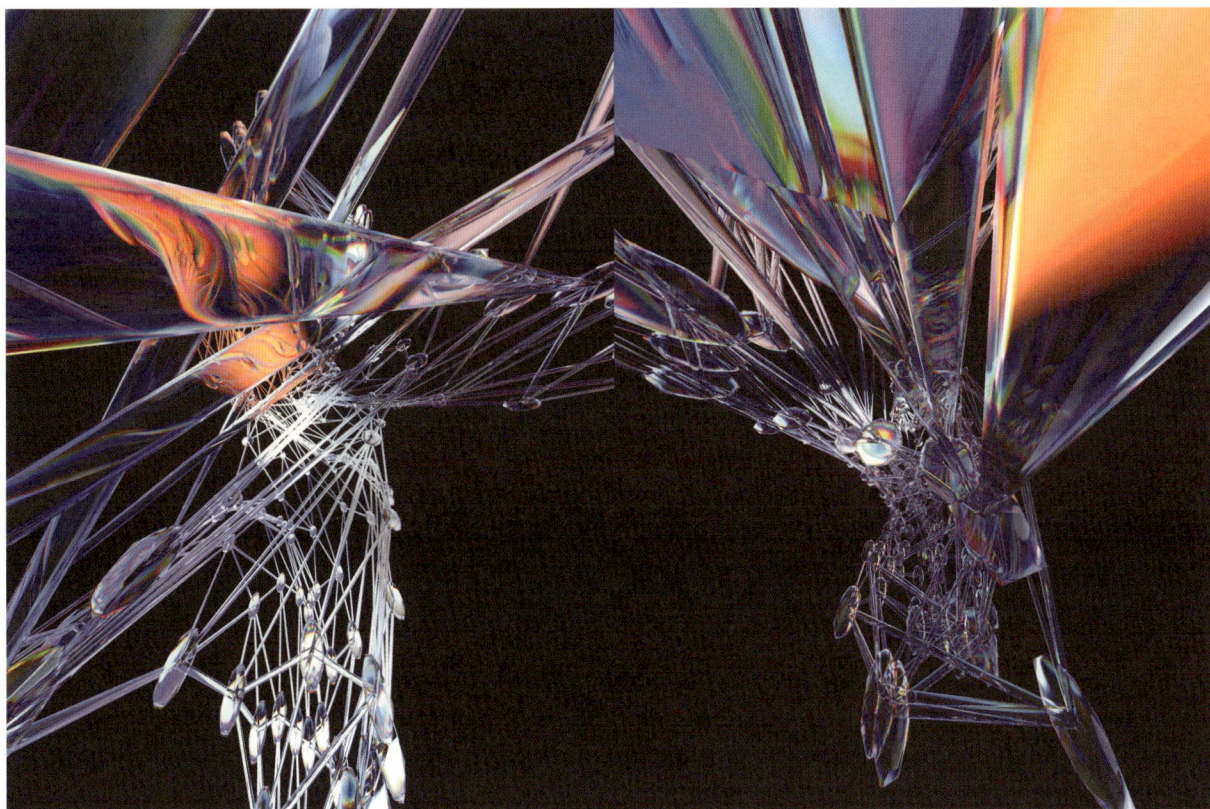

2.2.8.2 以《牡丹亭》中女性生命意识表达为背景下的虚拟可穿戴时尚拟态研究

2.2.8.2 Research on Virtual Wearable Fashion Mimicry in the Context of Female Life Consciousness Expression in *Peony Pavilion*

作者：杨甜、陈姝言、刘晓辰、胡玉暄、王涵、吴黎微

By Tian Yang, Shuyan Chen, Xiaochen Liu, Yuxuan Hu, Han Wang, Liwei Wu

产品以昆曲《牡丹亭》为背景，将其划分为"现实""梦境""冥界"三个场景，以其中植物的意象为线索，通过虚拟可穿戴产品展现梦与现实交融的生命存在状态，构建出其中虚拟的景象和女性生命意识。产品将《牡丹亭》作为拟态研究对象，进而将形式想象和物质想象相互配合，达到传统文化和数字虚拟的孪生关系的转化，并营造出从传统文化到当代社会的时空跨越，同时展现了从古至今，女性勇敢追求自由与个性解放的品质。

以代表美丽和端庄的花朵"牡丹"和象征自由与浪漫的"蝴蝶"为灵感，表现了丽娘渴望冲破礼制的束缚和渴望浪漫美好爱情的心理，呈现少女情怀及在梦境中与书生相恋缠绵的画面。

The product takes the Kunqu Opera *Peony Pavilion* as the background, divides it into three scenes of "reality" "dream" and "underworld", takes the imagery of plants as clues, and shows the life existence state of dream and reality through virtual wearable products, and constructs a virtual scene and female life consciousness. The product takes *Peony Pavilion* as the object of mimetic research. The cooperation of formal imagination and material imagination transforms the relationship between traditional culture and digital virtuality. It creates a time and space spanning from traditional culture to contemporary society while expressing the quality of women's courageous pursuit of freedom and personal liberation from ancient times to the present.

Inspired by the "peony", a flower representing beauty and dignity, and the "butterfly", a symbol of freedom and romance, the painting depicts Li Niang's desire to break free from the confines of propriety and her desire for romantic love. It reflects the feelings of a young girl and her dream of falling in love with a scholar.

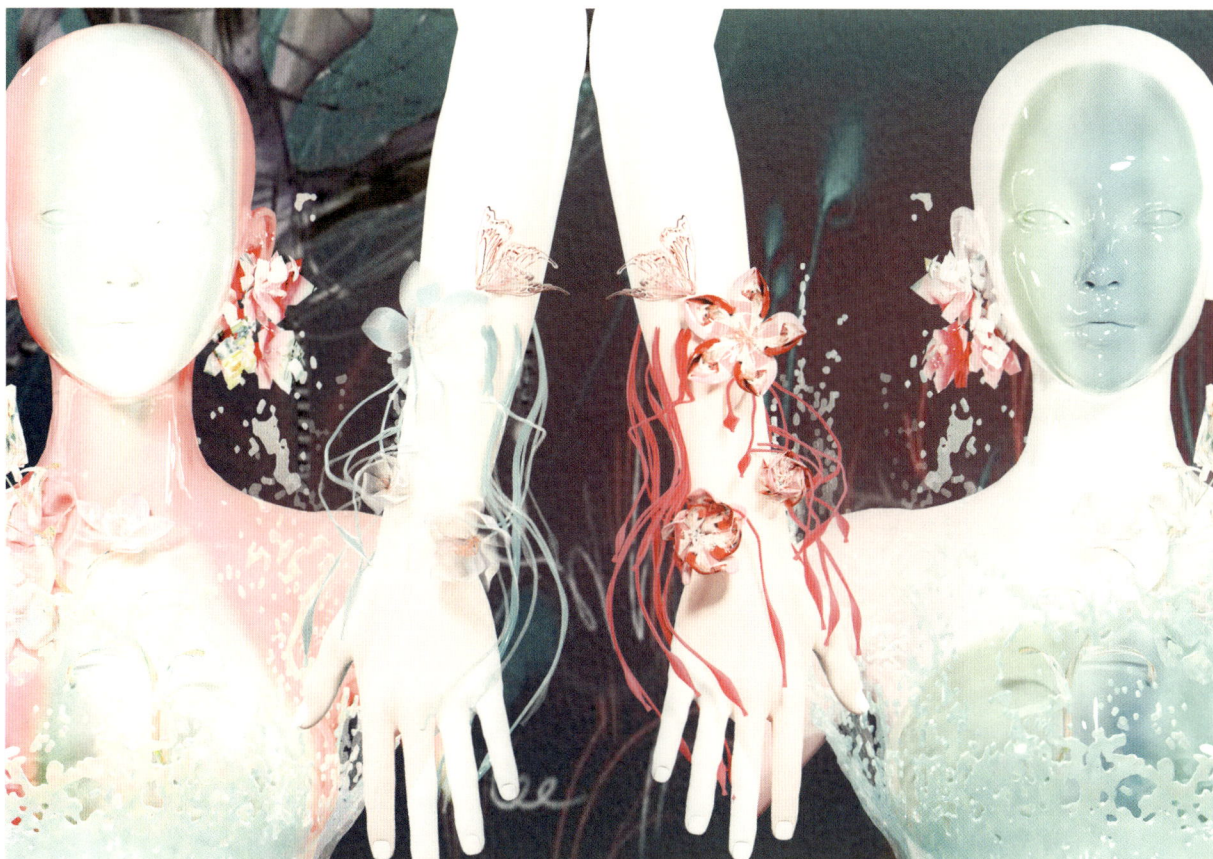

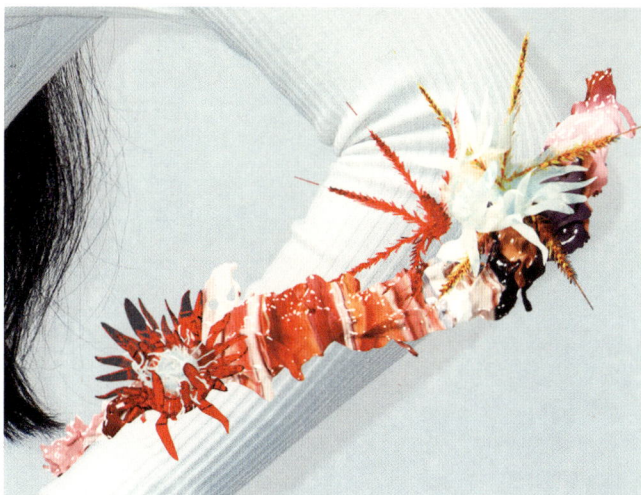

↘ **2.2.9 课题：幸福感成长桌游**

2.2.9 TASK: GROWING HAPPINESS BOARD GAME

课题说明：

　　调研表明，大学生心理健康问题呈逐年上升的趋势。大学阶段，是离开原生家庭，重塑自我的关键阶段，这期间的大学生面临着身份、学习方式、生活环境、交往群体等的转变，同时还会面临各种社会比较，是建立新的人生目标，树立人生观、价值观的关键时期。因此，了解大学生的心理健康状况，改进心理健康知识宣传方式，提升大学生自我训练和应对能力，是预防并减少危机事件的重要环节。据此，如何培养大学生正确的人生观，促进大学生幸福感的体验和获得人生长久的支撑是大学生成人成才的基石，"幸福观成长桌游"正是围绕这一问题的主动思考和回应。本实践研究在调查大学生心理健康现状和规律的基础上，帮助其感知、了解、面对自己的情绪状态，突破心理困境，建立自己稳定的长期幸福感支点。

指导老师：

　　宁兵，北京服装学院艺术设计学院数字媒体艺术系副教授
　　黄雅梅，心理学博士，北京服装学院心理健康教育与咨询辅导中心负责人

Task Description:

　　Research shows that mental health problems among university students are on the rise yearly. During this period, university students face changes in their identity, learning style, living environment, interaction groups, and so on, as well as various social comparisons, which are crucial for establishing new life goals and setting up outlooks and values. Therefore, understanding the mental health condition of university students, improving the way of promoting mental health knowledge, and enhancing their self-training and coping ability are essential links to prevent and reduce crisis events. Accordingly, cultivating a correct outlook on life, promoting the experience of happiness, and obtaining long-term support in life is the cornerstone of university students' adult success, and the "Growing Up with Happiness Board Game" is proactive thinking and response around this issue. This practical study is based on a survey of university students' current state and mental health patterns so that they can perceive, understand, and face their emotional state, break out of their psychological dilemmas, and establish their stable long-term fulcrum of happiness.

Tutor:

　　Bing Ning, Associate Professor of Digital Media Art, School of Art and Design, Beijing Institute of FashionTechnology
　　Yamei Huang, Ph.D. in Psychology, Dean of the Mental Health Education and Counseling Center of Beijing Institute of Fashion Technology

2.2.9.1 黎明之钥

2.2.9.1 The Key to Dawn

作者：高洁、龚怀谷、李飞跃、李思锦、马雅坤、马玥

By Jie Gao, Huaigu Gong, Feiyue Li, Sijin Li, Yakun Ma, Yue Ma

当今年轻人面临一个充满危机和不确定性的世界。为应对我们这一代人独特的挑战，桌游"黎明之钥"以积极心理学为蓝本构筑游戏框架，18天为一个周期，借助第一人称视角进行虚拟叙事，希望唤醒每位探险者心底沉睡的钥匙，迎来黎明破晓时。盒体轻便易携带，单人可游玩，没有时间空间的限制，只要你愿意，我们随时欢迎你踏上这趟旅程。我们希望每个人都能获得相对应的支持和治疗。

Young people today face a world of crisis and uncertainty. In response to the unique challenges of our generation, the board game "The Key to Dawn" is framed in positive psychology, with an eighteen-day cycle and a virtual narrative that draws on a first-person perspective, in the hope of awakening the dormant key to each explorer's heart and ushering in the dawn when it breaks. We hope that everyone can receive the corresponding support and treatment. The box is light and portable and can be played by a single person with no time or space constraints, so you are welcome to embark on this journey whenever you wish.

2.2.9.2　卷王 or 摆王

2.2.9.2　Overachiever or Laid Backer

作者：秦皓月、邵灵雁、宋静茹、王贞筑、朱欣悦、王艺蓓

By Haoyue Qin, Lingyan Shao, Jingru Song, Zhenzhu Wang, Xinyue Zhu, Yibei Wang

本桌游为 2～6 人的博弈棋盘游戏。在游戏中作为一名刚步入大学的大学生，面对自由、丰富的大学生活，你是选择成为智慧满分的"卷王"，还是享受生活的"摆王"呢？玩家可以根据自己对幸福校园生活的理解，探索地图，并完成地点发放的任务。但是无论做出怎样的选择，玩家都需要平衡好自己的各项数据：在成为"卷王"的路上不要忘记舒缓压力，在快乐开摆的路上也不要落下学习。我们想让玩家认识到，"平衡"的校园生活才是最幸福的。

This board game is a chess game for 2-6 people. In the game, as a college student just entering the university, facing the free and rich university life, do you choose to be the Overachiever with full wisdom or the Laid backer? Players can explore the map according to their understanding of happy campus life and complete the task of distributing the location. However, no matter your choice, players need to balance their data: don't forget to relieve pressure on the way to becoming the Overachiever, and don't throw away learning on the way to becoming a Laid backer. We want to convey to the players that "balanced" campus life is the happiest.

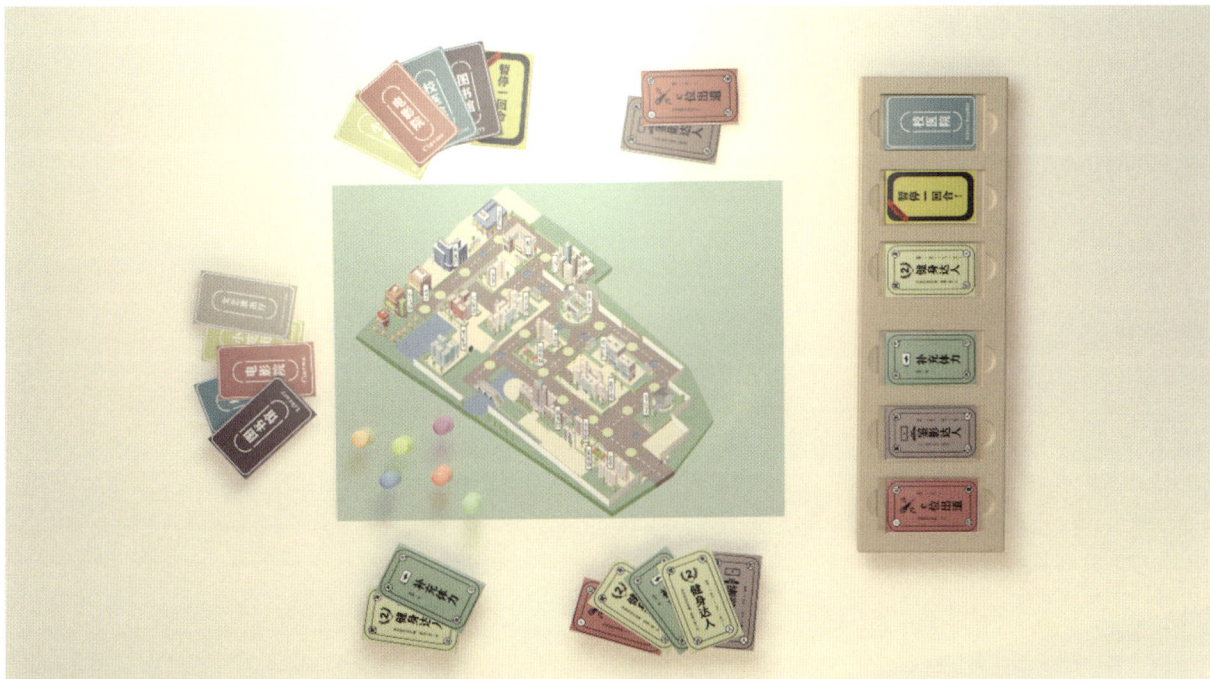

↘ **2.2.10 课题：小确幸大生活 —— 定格动画设计课程**

2.2.10 TASK: SMALL FORTUNES, BIG LIVES—STOP-MOTION ANIMATION DESIGN COURSE

课题说明：

    时间从未停滞匆忙的脚步，世界始终保持高速运转，每一个生命个体都在面临着不同程度的压力与考验。导师组成员希望每一个参与工作坊的成员，都将各自人生中经历的值得追忆的生活片段、那些震撼或打动我们的瞬间，那些在他人看来平淡无奇，但对我们意义重大的小事……通过定格动画的语言制作一份汇集众人生活的记忆群像，纪念我们悄然逝去的青葱岁月，寻找内心深处的平静与自在。保留内心深处的"优雅"，才能张开双臂更好地拥抱未来。

指导老师：

    余春娜，教授，硕士生导师，天津美术学院实验艺术学院动画艺术系主任

    张轶，天津美术学院实验艺术学院动画艺术系副教授

    索璐，天津美术学院实验艺术学院动画艺术系讲师

    张玥，天津美术学院实验艺术学院动画艺术系讲师

Task Description:

Time never stops its hurried pace, and the world keeps moving at a high speed. Individual life is under varying degrees of stress and testing. The mentorship team members hope each workshop participant will present the memorable moments of their lives that shocked or touched us, the small things that seemed ordinary to others but meant a lot to us. Through the language of stop-motion animation, we have created a collection of memories of our lives, commemorating our lost years and finding peace and ease in our hearts. When we pick up these "small fortunes" one by one, we will have found inner peace and ease, preserving the "elegance" deep inside us so we can embrace the future with open arms.

Tutor:

Chunna Yu, Professor, Master's Supervisor, Dean of Department of Animation Art, School of Experimental Art, Tianjin Academy of Fine Arts

Yi Zhang, Associate Professor, Department of Animation Art, School of Experimental Art, Tianjin Academy of Fine Arts

Lu Suo, Lecturer of Department of Animation Art, School of Experimental Art, Tianjin Academy of Fine Arts

Yue Zhang, Lecturer of Department of Animation Art, School of Experimental Art, Tianjin Academy of Fine Arts

2.2.10.1　千人千眼，千眼千幸

2.2.10.1　Thousands of Eyes，Thousands of Happiness

作者：张鹤野、高齐子真、尹东月

By Heye Zhang, Qizi Zhen Gao, Dongyue Yin

　　"小确幸"成为当下人们对美好生活的期望。但我们所希望的"小确幸"，其实就藏匿在生活的点滴之中，并构筑着生活。我们通过眼睛观察并接收生活的信号，一千个人眼中有一千个哈姆雷特。动画"千人千眼，千眼千幸"串联了来自不同"眼睛"所观察到的"小确幸"片段，通过定格动画的语言组成一份汇集众人生活的记忆群像，纪念终将成为过去的当下，由此寻找内心深处的平静与自在。

　　The "small blessings" are now people's expectations for a better life. But the "small blessings" we hope for are hidden in the little things that make up our lives. We see and receive the signals of life through our eyes. There are a thousand Hamlets in the eyes of a thousand people. The animation "Thousands of Eyes, Thousands of Happiness" connects the "small true happiness" fragments observed from different "eyes". The language of stop-motion animation forms a group image of the life of all people, commemorating the present that we will eventually become the past and looking for peace and freedom in our hearts.

2.2.10.2　小幸福和大生活

2.2.10.2　Small Happiness and Big Life

作者：刘师然、李萌、李培瑜、萧扬飞、王雪

By Shiran Liu, Meng Li, Peiyu Li, Yangfei Xiao, Xue Wang

在人生的道路上我们都需要停下脚步，尝试寻找生活中的小美好，不要辜负生活的意义。这些生活中的幸运，或许是心灵的安慰、精神的满足、身体的饱腹，又或者是一见钟情的喜悦。有些时候，我们或许能够停下繁忙的脚步，回头看看生活中的那些美好，很可能因此发现许多自己从未注意到的乐趣，或者是重拾对生活的兴趣。

We all need to stop on the road of life, try to find the petite beauty in life, and not live up to the meaning of life. These lucky things may be the comfort of the heart, the satisfaction of the mind, the fullness of the body, or the joy of falling in love at first sight. Sometimes, we can stop our busy pace and look back at the good things in life. You may discover many pleasures that you never knew existed, or you may rekindle your interest in life.

↘ **2.2.11　课题：以身体之——具身性设计课程**

2.2.11　TASK: DESIGNING WITH THE BODY—EMBODIED DESIGN COURSE

课题说明：

当我们越来越依赖于虚拟交流，我们的行为、思维与未来也发生了改变。我们通过唤醒身体来思考：身体是否有可能塑造精神？身体如何与自然交互？身体如何响应环境？可否从身体感知出发获得心智力量？都是我们将要回答的问题。课程以开放的心理学、现象学、生物科学、人体美学、工程学和材料学等跨学科的启发与感知为起点，开展对于身体观、身体的问题意识、身体的未来等问题的研究与解读。课程颠覆传统从理论到实践的认知模式，引导学生通过"体"得到"验"，从哲学、医学、音乐、戏剧等工作坊中得到启发，学会用身体感知世界，重新建立与世界的联系。把身体当作一块无限延伸的巨大画布，探索身体与物、身体与场、身体与自然、身体与材料之间的关系，启动身体计划，催化身体反思，推演与展望设计的全新趋势与人类生态的未来。

Task Description:

As we become increasingly dependent on virtual communication, our behavior, thinking and our future change. Through awakening the body, we think: is it possible for the body to shape the spirit? How does the body interact with nature? How does the body respond to its environment? Is it possible to gain the power of the mind from physical perception? These are the questions we will answer-Starting from an interdisciplinary perspective that is open to inspiration and perception from psychology, phenomenology, bioscience, body aesthetics, engineering and materials science, the course introduces embodied cognitive theory. It explores new perspectives on the body, the body's problematic awareness, and the future of the body. The course overturns the traditional theory-to-practice cognitive model, guiding students to "test" through the "body", to be inspired by philosophy, medicine, music, theater, and other workshops, to learn to use the body to perceive the world and to re-establish a connection with the world. Students will learn to use their bodies to perceive the world and re-establish their connection with the world. By treating the body as a vast canvas with infinite extension, they will explore the relationship between the body and objects, the body and fields, the body and nature, and the body and materials.

指导老师：

向逸，中国美术学院教师
马川，中国美术学院创新设计学院技术与造物研究所副所长
甘道夫，认知科学与艺术表达教师
赵雨，博士，西南大学设计学副教授，硕士生导师

Tutor:

Yi Xiang a teacher at China Academy of Art
Chuan Ma, Deputy Director of the Institute of Technology and Creation, School of Innovative Design, China Academy of Art
Gandalf, a teacher of cognitive science and artistic expression
Yu Zhao, Ph.D., Associate Professor, Master's Supervisor, Southwest University

产妇关怀 VR

Postpartum Care VR

作者：乜可昕、王润琪、周国宝、朱韩剑、邵文茜、刘欣然

By Kexin Nie, Runqi Wang, Guobao Zhou, Hanjian Zhu, Wenxi Shao, Xinran Liu

在 VR 产后心理关怀设计方案中，我们设计了四个模式供用户选择，并在一些模式中设置了 VR 场景互动动作引导，将轻运动与各个场景融合在一起，帮助产后女性在家中放松心情的同时也可以进行一些简单的锻炼。在单人空间的模式下设有不同的场景入口，如冥想和私人影院场景，给产后女性提供一定的个人休息空间；二人空间的模式中包含情侣约会场景、私人影院等，帮助产后女性加强与丈夫的情感交流；亲子空间模式可以上传孩子的日常动态，满足产后妈妈喜欢"晒娃"的分享欲；在社会空间中，产后女性可以通过元宇宙 VR 社群与其他产妇建立联系，帮助产后女性尽快与社会接轨。

Our VR postpartum care design proposal focuses on releasing psychological pressure for postpartum people. We designed our modes for users to choose from and set up the various embodied interactive actions in different VR scenes. At the same time, integrating light sports with several scenes helps postpartum people relax at home. Different environments, such as meditation and private theater, are provided in the personal space mode. The couple mode focuses on the dating area that helps postpartum people enhance emotional communication with their husbands. Users could upload their children's daily lives within the parent-child space to satisfy their desire of sharing babies. The society space offers an opportunity to connect with other mothers through the Metaverse VR community to promote postpartum people to integrate with real-life society as soon as possible.

2.2.12 课题：增进获得：为亚洲用户的无障碍日常生活而设计

2.2.12 TASK: ENHANCING ACCESS: DESIGNED FOR THE ACCESSIBLE EVERYDAY LIFE OF ASIAN USERS

课题说明：

越来越多的设计师达成了为弱势群体设计的共识，无障碍设计已应用于生活场景的诸多方面。无障碍设计是针对身体能力差异，做出对残障使用场景友善的设计，此类设计关注残疾人、老年人的特殊需求，但并非只是为残疾人、老年人群体设计，人们在特殊情境下会发生能力的巨大变化，产生临时性和情境性残障，比如占用视觉注意资源的驾驶过程、手部负伤的家务劳动等。优秀的无障碍设计不仅能让特殊群体用户正常地与产品交互，还能够为普通人的临时性障碍情景提供更好的使用体验。为永久性残疾人设计似乎是一个很受限制的方向，但由此产生的设计实际上可以使更多的人受益。

指导老师：

董理权，中国残疾人辅助器具中心副主任
山娜，北京服装学院服饰艺术与工程学院工业设计专业助理教授
张帆，北京服装学院服饰艺术与工程学院工业设计专业助理教授
崔艺铭，北京服装学院服饰艺术与工程学院产品设计专业助理教授

Task Description:

As more and more designers develop a consensus to design for the disadvantaged, accessible design has been applied to many aspects of life scenarios. Accessible design is a disability-friendly design that addresses the differences in physical abilities and focuses on the unique needs of people with disabilities and older people, but not just for them. People can experience significant changes in their abilities in particular situations, resulting in temporary and situational disabilities, such as driving that takes up visual attention resources or domestic tasks that involve hand injuries. Good accessibility design allows special groups of users to interact with products normally and provides a better experience for ordinary people with temporary impairment situations. Designing for people with permanent disabilities may seem like a significant limitation, but the resulting designs can benefit many more people.

Tutor:

Liquan Dong, Deputy Dean of the China Center for Assistive Devices for the Disabled
Na Shan, Assistant Professor, Industrial Design, School of Fashion Art and Engineering, Beijing Institute of Fashion Technology
Fan Zhang, Assistant Professor, Industrial Design, School of Fashion Art and Engineering, Beijing Institute of Fashion Technology
Yiming Cui, Assistant Professor, Industrial Design, School of Fashion Art and Engineering, Beijing Institute of Fashion Technology

2.2.12.1 针对东南亚手部功能性障碍患者日常生活辅助器具设计

2.2.12.1 For Patients with Hand Dysfunction in Southeast Asia Assistive Devicedesign for Daily Life

作者：陈琳、李赢、牛禹丹、刘思聪、徐玉晗、秦朗

By Lin Chen, Ying Li, Yudan Niu, Sicong Liu, Yuhan Xu, Lang Qin

项目方案旨在通过弥补手部运动功能的不足，打破隔离残疾人士与优雅生活的障碍。产品通过A+B与A+C模块多元组合帮助手部运动功能障碍患者实现拧、握、拉等生活常用动作。A模块固定于中指，B、C模块利用掌心异形结构填充患者握不拢拳头（抓取动作）时掌心的空缺。希望能利用轻巧便捷的手部功能辅助器在各种社会生活领域为残疾人的平等参与做出小小的贡献。

The project aims to break down the barriers that isolate people with disabilities and elegant living by making up for the sound motor function of the hands. The product helps patients with hand movement dysfunction to achieve everyday life movements such as twisting, gripping, and pulling through the multiple combinations of A+B and A+C modules. The A module is infixed to the middle finger, and the B or C module uses a palm-shaped structure to fill the gap in the palm's palm when the patient does not hold the fist (grasping action). Using a lightweight and convenient hand function aids in making a small effort to promote the equal participation of persons with disabilities in all areas of social life.

2022 DESIGN DAY 设计马拉松 **针对手部运动功能障碍用户的日常生活辅助器（模块间可多元搭配组合）**
Daily life aids for users with hand movement dysfunction (multiple combinations can be combined between modules)

A+C

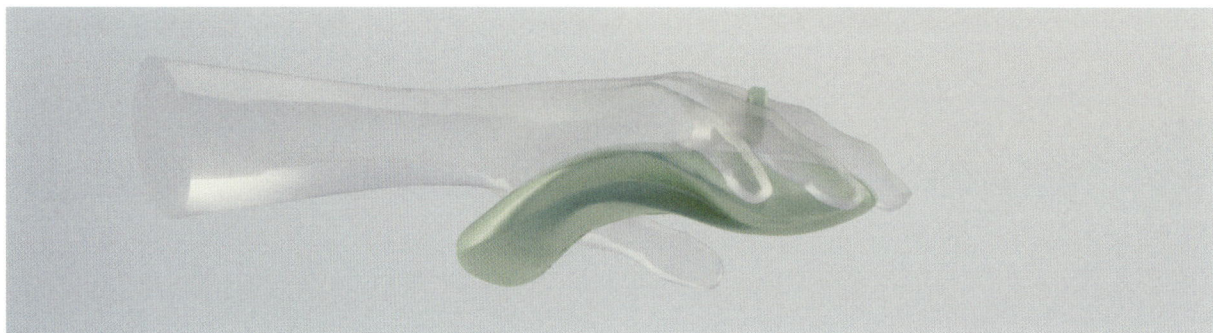

针对手部运动功能障碍用户的日常生活辅助器（模块间可多元搭配组合）

Daily life aids for users with hand movement dysfunction (multiple combinations can be combined between modules)

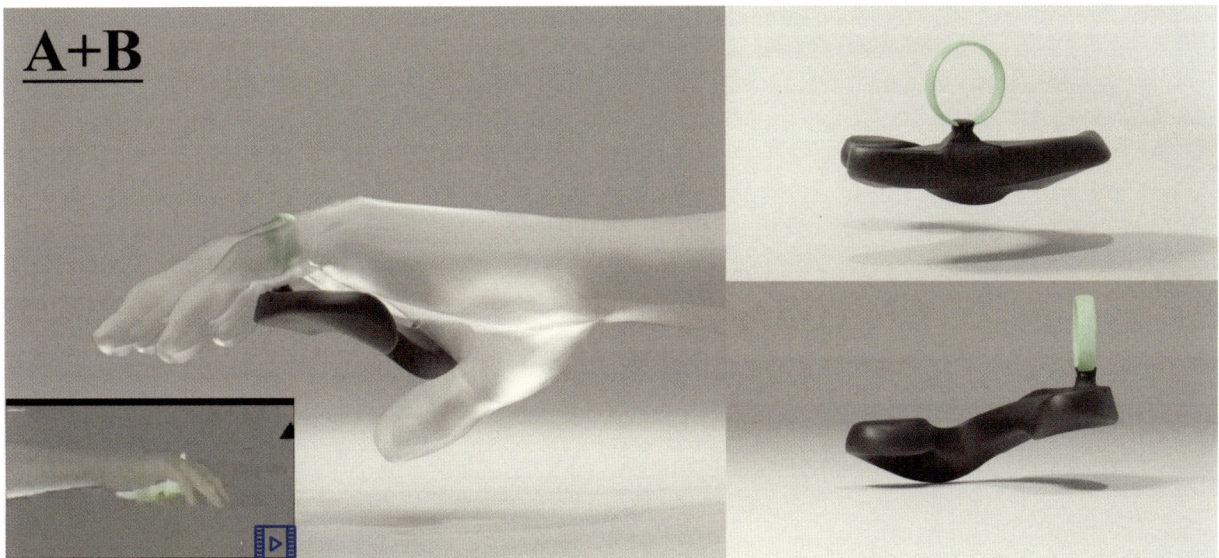

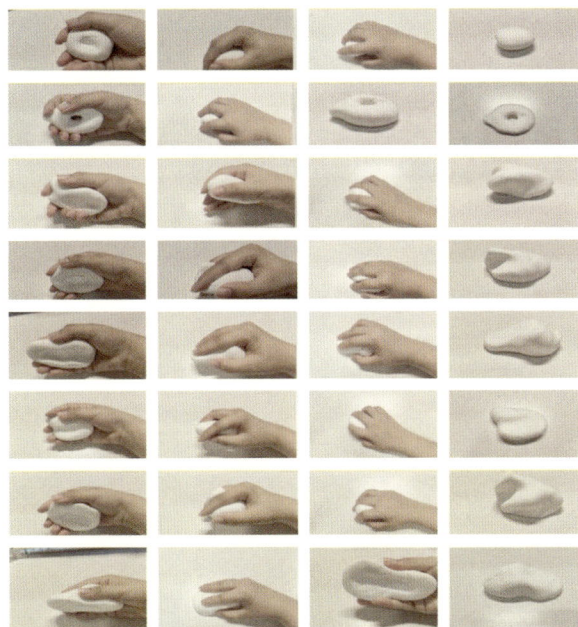

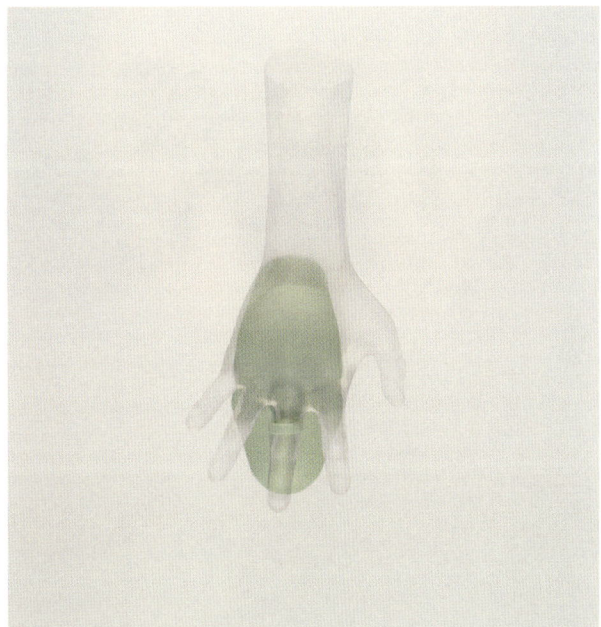

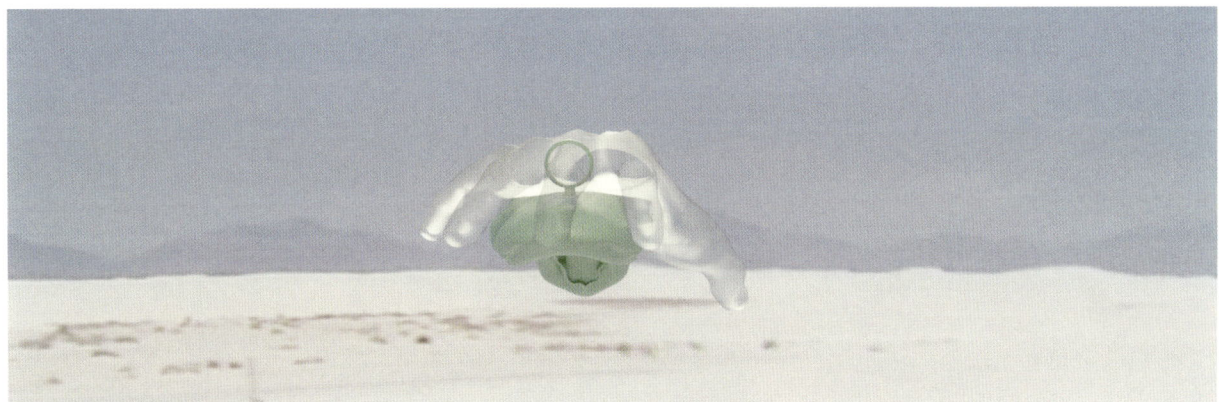

2.2.12.2 针对东南亚无障碍群体的临时安置居所设计

作者：史方正、贺盈颖、王温雅、权雯欣、谢佳薇、陈漫霏

东南亚地区有亚热带季风性气候的特殊性，洪旱灾害是东南亚地区较为频发的自然灾害，自1970年以来，亚洲地区和太平洋沿岸地区的自然灾害影响了69亿人。我们聚焦灾后快速重建，以及无障碍群体的使用需求。本次设计基于东南亚国家的无障碍临时安置居所，采用盒子作为模块搭建的基本设想，结合东南亚地区特有的房屋结构和本土材料进行预制模块的搭建，并逐步完善关于房屋规格及元件拼接的进一步设想。本设计结合了东南亚地区干栏建筑的特点以保持稳定性与疏水透风性，并基于此设定了居住区、淋浴区、活动间、卫生间等不同的使用空间。

2.2.12.2 Design of Temporary Housing for Barrier-free Groups in Southeast Asia

By Fangzheng Shi, Yingying He, Wenya Wang, Wenxin Quan, Jiawei Xie, Manfei Chen

Southeast Asia has an exceptional subtropical monsoon climate. Floods and droughts are more frequent natural disasters in Southeast Asia, which have affected 6.9 billion people in Asia-Pacific region since 1970. Based on the design of accessible temporary housing in Southeast Asian countries, we focus on the rapid reconstruction after the disaster and the needs of the accessible groups to use. This design uses a box as the module construction concept and combines Southeast Asia's house structure and local materials to build prefabricated modules. Further ideas about the house specifications and component splicing are refined. Based on the characteristics of dry-rail construction in Southeast Asia to maintain stability and air permeability, different spaces such as living areas, shower areas, activity rooms, and bathrooms were set.

↘ 2.2.13 课题：15 分钟生活圈："新常态"下的人居环境设计

2.2.13 TASK: 15-MINUTE LIVING CIRCLE: RESIDENTIAL ENVIRONMENT DESIGN UNDER THE "NEW NORMAL"

课题说明：

"15分钟生活圈"是一个都市规划概念，其目标是让城市社区的居民都可以在步行或脚踏车路程可及的范围内，满足衣食住行娱乐等大部分日常所需的服务。这个概念最初是由法裔哥伦比亚学者卡洛斯·莫雷诺（Carlos Moreno）于2016年提出，随后由法国巴黎市市长安妮·伊达尔戈（Anne Hidalgo）推广。这个概念可以被描述为"回归当地的生活方式"。本次工作坊的愿景是重新定义健康、安全的美好人居环境空间，设计的对象是基于每个人身边的15分钟环境空间。我们将基于"设计四秩序"（符号—物体—行动—想法）的模型来思考人居环境的系统性解决方案。区别于传统教学设计"结果导向型"的教学方式，更强调设计"思维系统性"的训练。在整个设计过程中以"双钻模型"（发现—定义—发展—交付）为导向，强化训练同学们的以人为本的、系统的、协同的和可视化的"设计思维"。

指导老师：

杨叶秋，意大利米兰理工大学设计学在读博士研究生，英国密德萨斯大学访问学者，同济大学设计创意学院访问学者

崔咏梅，上海立达学院艺术设计学院助理教授

陈宇轩，英国格拉斯哥艺术学院创新与服务设计系硕士

程楚涵，意大利米兰理工大学产品服务系统设计系硕士

楼振罡，意大利米兰理工大学产品服务系统设计系硕士

刘帅，意大利米兰理工大学设计与工程系硕士

李天翼，意大利米兰理工大学产品服务系统设计系硕士

Task Description:

The "15-minute living circle" is an urban planning concept. It's goal is to enable residents of urban communities to access most of their daily needs, including clothing, food, housing, transportation, and entertainment, within walking or biking distance. This concept was proposed by Carlos Moreno, a French-Colombian scholar, in 2016 and then promoted by Anne Hidalgo, mayor of Paris. This concept can be described as "returning to the local way of life". The vision of this workshop is redefining a healthy, safe, and beautiful residential space. The design object is based on the 15-minute environmental space around everyone. We will think about the systematic solution of human settlement environment based on the model of "Four Orders of Design" (symbol-object-action-idea). Unlike the traditional teaching method of "result-oriented" design, we emphasize training in "systematic thinking" in design. Throughout the design process, the "Double Diamond Model" (discover-define-develop-deliver) is used as a guide to training students to develop human-centered, systematic, collaborative, and visualized "design thinking".

Tutor:

Yeqiu Yang, a Ph.D. candidate in Design, Polytechnic University of Milan, Italy; visiting scholar at Middlesex University, UK; visiting scholar at the School of Design and Creativity, Tongji University

Yongmei Cui, Assistant Professor, School of Art and Design, Shanghai Lida University

Yuxuan Chen, Master of Innovation and Service Design, The Glasgow School of Art, UK

Chuhan Cheng, Master of Product Service System Design, Polytechnic University of Milan, Italy

Zhengang Lou, Master of Product Service System Design, Polytechnic University of Milan, Italy

Shuai Liu, Master of Design and Engineering, Polytechnic University of Milan, Italy

Tianyi Li, Master of Product Service System Design, Polytechnic University of Milan, Italy

2.2.13.1 缝合社区计划

2.2.13.1 The Stitching Community Project

作者：张梦蝶、郭子琦、周圆圆、白沐凡、陈颖颖、贝莎莎

By Mengdie Zhang, Ziqi Guo, Yuanyuan Zhou, Mufan Bai, Yingying Chen, Shasha Bei

我们的项目名称为社区缝合计划。我们设计的15分钟生活圈位于天津市风光里社区。这个社区是一个有四十年历史的老旧社区，近四十年来几乎没有被改造更新过，因此我们一开始设定的目标就是：及时有效、低成本地改善此地区的老年居民的生活。为此我们设计了社区智能模块服务系统，将扩音器、无线充电、手电筒这样的生活工具集中到一个方盒子中，同时也与周围的商店和医院建立了快速联系渠道，并将这些智能工具放置在社区人流聚集点，帮助老年人快速阅读获取信息，同时辅助购药和购物。视觉上我们使用了具有识别度的高饱和色彩，同时使用简单的交互方式吸引老年人使用。

Our project is called the Community Stitching Project. Our designed 15-minute living circle is in the scenic neighborhood of Tianjin, China. This neighborhood is a forty-year-old neighborhood. It has hardly been remodeled in nearly four decades, so we set the goal at the outset of improving the lives of the elderly in the area in a timely, effective, and low-cost manner. To this end, we have designed a community intelligent module service system; we will be amplifiers, wireless charging, flashlights, and other life tools in one square box. Quick contact tools have also been set up with surrounding shops and hospitals. Placing these smart tools at community gathering points helps seniors read quickly for information while assisting in drug purchases and shopping. Visually, we use highly saturated colors with recognition and simple interactive methods to attract the elderly.

2.2.13.2 "共生·自然"校园生态空间服务系统设计——以南京艺术学院孔雀园为例

2.2.13.2 "Symbiosis and Nature" Campus Eco-Space Service System Design—The Peacock Garden of Nanjing University of the Arts as an Example

作者：刘远航、秦鹏、沈浩书、吴莹小、马恺月、吴富钢

By Yuanhang Liu, Peng Qin, Haoshu Shen, Yingxiao Wu, Kaiyue Ma, Fugang Wu

　　我们使用双钻设计模型主导设计全流程，将问题聚焦于人与自然的互动关系上。借助"设计四秩序"，在第一领域，提取孔雀相关元素，产出孔雀园视觉形象及衍生平面；第二领域，设计校园内空间互动装置；第三领域，我们以建立孔雀园服务体系为目标，结合线上功能，设计空间游戏和其中的体验；第四领域，我们将整合打造能够进行知识宣传、线上预约、同时获得多方支持的自然共生的场景。在校园静谧之处，创新人与孔雀自然共生的互动关系，营造一个服务于校园生态空间的场所。

　　We use the Double Diamond design model to lead the design process, focusing on the interaction between people and nature. With the help of the "Four Orders of Design", in the first area, we extracted peacock-related elements to produce the visual image of the Peacock Garden and its derivative planes. In the second area, we designed an interactive installation on the campus. In the third area, we aim to establish the service system of Peacock Garden and design spatial games and experiences in combination with online functions. In the fourth area, we integrated to create a natural symbiosis scene capable of knowledge promotion, online reservation, and simultaneous support from multiple parties. In the quiet place of the campus, we innovate the interaction between people and peacocks in natural symbiosis and create a place that serves the ecological space of the campus.

2.2.13.3　从忽略到反哺——针对中老年群体的收发快递问题的便民服务设施改造

2.2.13.3　The Transformation of Convenient Service Facilities—Aimed at the Delivery Problems of Middle-Aged and Elderly Groups Ignored by Society

作者：杜宙飞、张诗涵、何伟、黎广贤、王家妤

方案以老年人快递收发问题为核心进行体系化服务设计，从多角度反哺关爱老年人，以多方案回应其被忽略的需求。项目选址位于北京市紫荆社区。设计的核心内容在于能够提供收发快递、派送无人车服务，同时高度与APP联动的快递站。衍生出配套的适老化设计内容，如APP设计、快递柜更新设计、无人车派放系统、快递垃圾回收系统等，从老人视角出发，解决老人的难题，同时提高社区资源利用率，反哺老旧社区。

By Zhuofei Du, Shihan Zhang, Wei He, Guangxian Li, Jiayu Wang

The programme takes the issue of sending and receiving couriers for older adults as the core of systematic service design, feeds and cares for them from multiple perspectives, and responds to the neglected needs with multiple ways. We selected the Beijing Zijing Community as an example. The core content of the design is to provide an express station with the services of sending and receiving express delivery, delivering unmanned vehicles, and highly linking with the app simultaneously. Derived supporting aging design contents, such as app design, updated design of the express cabinet, unmanned vehicle delivery system, express garbage collection system, etc., to solve the problems of the elderly from the perspective of the elderly, while improving the utilization rate of community resources and feeding back the old communities.

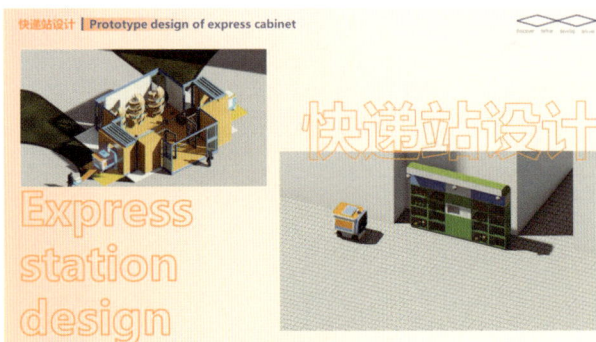

快递站设计 | Prototype design of express cabinet

Express station design

快递站设计 | Prototype design of express cabinet

应用交互界面 | Application interactive page

无人车快递车功能

应用交互界面 | Application interactive page

老年人餐食物资购物端

↘ 2.2.14 课题：优雅地退休——包容性设计赋能老龄产品创新

2.2.14 TASK: RETIRE GRACEFULLY—INCLUSIVE DESIGN EMPOWERS INNOVATION IN AGING PRODUCTS

课题说明：

包容性设计是"一种不需适应或特别设计，而使主流产品或服务能被尽可能多的用户所使用的设计方法和过程"。人口老龄化日益严重，构建一个适合老人生活的社会，是我们每一个人都可以参与、需要为之努力的事情。老年人的身体机能随着年龄增大，或多或少会有所衰退，而老年人使用产品与服务的能力和其自身的能力（比如感官能力——视觉和听觉；运动能力——移动、伸展和灵活性；认知能力——交流和智力）密切相关。本课题从跨感官补偿的角度切入，赋能（补偿、提升用户能力）老年用户，提升产品使用体验。工作坊中会介绍包容性服务设计的相关工具和方法，以及英国剑桥大学工程设计中心关于包容性设计的研究。希望与工作坊的学员一起通过设计手段来创新老龄产品、服务，让更多的人了解和关注老龄人群的需求，助力老年人优雅地退休与生活。

指导老师：

刘源源，北京航空航天大学机械工程及自动化学院工业设计系副教授，硕士生导师

Task Description:

Inclusive Design is "a design method and process that makes mainstream products or services available to as many users as possible without adaptation or special design. "With an increasingly aging population, building a society fit for older adults is something we can participate in and need to work towards. The use of products and services by older people is closely related to their abilities (e.g., sensory abilities—sight and hearing, motor abilities—movement, stretching and dexterity, and cognitive abilities—communication and intelligence), as their physical functions deteriorate to a greater or lesser extent as they age. This task is approached from the perspective of cross-sensory compensation to empower (compensate or enhance the user's abilities) older users and enhance the product experience. The workshop will introduce tools and methods related to inclusive service design and research on inclusive design from the Centre for Engineering Design at the University of Cambridge, UK. We hope to work with workshop participants to innovate products or services through design tools so that more people will understand and pay attention to the aging population's needs and help older people retire and live gracefully.

Tutor:

Yuanyuan Liu, Associate Professor, Master's Supervisor, Industrial Design Department, School of Mechanical Engineering and Automation, Beihang University

2.2.14.1 伴熟食堂：代际互动社区共享膳食计划

2.2.14.1 Accompanying Cooked Canteen: Intergenerational Interactive Community Shared Meal Plan

作者：何丽雯、黄雨晴、刘佳慧、宋雨露、Chaeyun Jung、蒋新

By Liwen He, Yuqing Huang, Jiahui Liu, Yulu Song, Chaeyun Jung, Xin Jiang

这是一个为实现退休人群与在外漂泊青年互动、互助、互相治愈的共享膳食服务设计。首先，它依托于线上订餐、线下取餐环节，鼓励用户以膳食分享进行上门交流，助力老青互动，建立友好型社区。此外，它还在节假日增加下厨、交流分享等环节，以线下生活小站为介，为用户提供一个愉悦的交流场所和膳食分享的空间。由此在社区中构建退休人群与漂泊青年之间沟通的桥梁。

This design is a shared meal service for enable interaction, mutual support and healing between retired people and young people who are away from home. Firstly, it relies on online meal ordering and offline meal pick-up sessions to encourage users to communicate at home through meal sharing, helping old and young people interact and build a friendly community. In addition, it also holds cooking and sharing sessions on holidays, with offline life stations as a mediator and a pleasant place for users to communicate and share meals. This builds a bridge between retired people and drifting youths in the community.

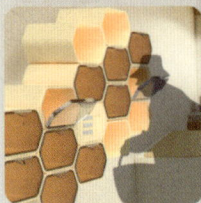

2.2.14.2 乐归园——打造环创、互助、愉悦的理想退休生活方式

2.2.14.2 Joyous Planed—Creating an Ideal Retirement Lifestyle of Environmental Creativity, Mutual Support and Pleasure

作者：李岩、岳乐、姜知爱、赵跃、唐煜霖、李良斌

By Yan Li, Le Yue, Zhiai Jiang, Yue Zhao, Yulin Tang, Liangbin Li

调研发现，大多数退休老人都会面临社交困境、身体机能减弱以及自我否定等问题，而种植行为是老年人较为广泛的爱好。"乐归园"是提供可移动的组合式种植模块，让退休老人日常耕种，在打造社区园林景观的同时，使老人们老有所为。此外，老年人可以为小区有植物寄养需求的人群，如上班族提供植物寄养服务。退休老人们可以在种植场所充分交流，提升他们的社交满足感，使退休老人老有所乐。

Through research, we found that most retired older adults will face social difficulties, weakened physical function, self-denial, and other problems, and planting behavior is a more widespread hobby of older adults. Joyous Planed is designed to provide the retired older adults with daily farming through the movable combined planting module so that older adults can do something while creating the community landscape. For plant foster care services, older adults can provide services for people, such as office workers with plant foster care needs in the community. In addition, retirees can fully communicate in the planting place to enhance their social satisfaction and make them happy.

↘ 2.2.15 课题：声音的感知与重构

2.2.15 TASK: SOUND PERCEPTION AND RECONSTRUCTION

课题说明：

聆听是人们生活中不可或缺的部分，我们通过听觉感知时间运行，从而了解世界，了解我们身处的环境和社会。现在，借助计算机程序和音频设备，我们能够根据自己的感知用声音重新构建认知中的世界，采集我们认为有趣、丰富的各类日常声音，结合不同类型的视觉媒介，创造出将现实和想象融为一体的视听世界。本课程将带领学员认识音乐、认识声音，并结合学员自己的专业基础知识，学习数字音频工作站的应用技术，掌握音乐创作的技巧，培养创作音乐和视听一体化作品的能力。

指导老师：

亓梦婕，中国音乐学院教师，中央音乐学院博士后研究员

Task Description:

Listening is an integral part of people's lives. We learn about the world, our environment, and society deep within us through our auditory perception of the orbit of time. Now, with the help of computer programs and audio devices, we can reconstruct the world as we perceive it with sound, capturing all kinds of everyday sounds that we find interesting and rich, and bringing together different types of visual media to create audio-visual worlds where reality and imagination merge. This course will take participants through an introduction to music, to sound, and to the application of digital audio workstations in conjunction with their professional grounding in the techniques of music composition and the development of the ability to create integrated musical and audio-visual works.

Tutor:

Mengjie Qi, Lecturer of China Conservatory of Music, Postdoctoral Researcher at the Central Conservatory of Music

2.2.15.1　以声音接近世界

2.2.15.1　Sound Close to The World

作者：彭汉婷、刘雨婷、王辰雨、郑妮娜、韩伟

By Hanting Peng, Yuting Liu, Chenyu Wang, Nina Zheng, Wei Han

　　我们正处在一个不稳定的临时状态之中，我们被嵌入在一个多变量系统当中无法脱身。在这个时代里，人的听觉被喧嚣掩盖，城市变得缺乏可听性，失去了对声音原始的掌控感。我们通过生成艺术的方式将声音可视化，并以装置的形式，展现当外部变量持续增加至无法掌握的局面时，造成的临时状态常态化。表现乌卡时代（VUCA）下，人群被大量时代信息反复、无意识地触发失控感，以此方式思辨人们在全球体系下的处境问题。

　　We are in an unstable temporary state. We are embedded in a multivariate system from which we cannot escape. In this era, people's hearing is covered up by the noise, the city has become less audible and has lost its original sense of control over the sound. We visualize the sound using generative art, and in the form of an installation, we show the normalization of the temporary state caused when the external variables continue to increase and reach a situation that cannot be grasped. In the VUCA era, the crowd's sense of loss of control is repeatedly and unconsciously provoked by a large amount of information about the era. In this way, people's situation in the global system is speculated.

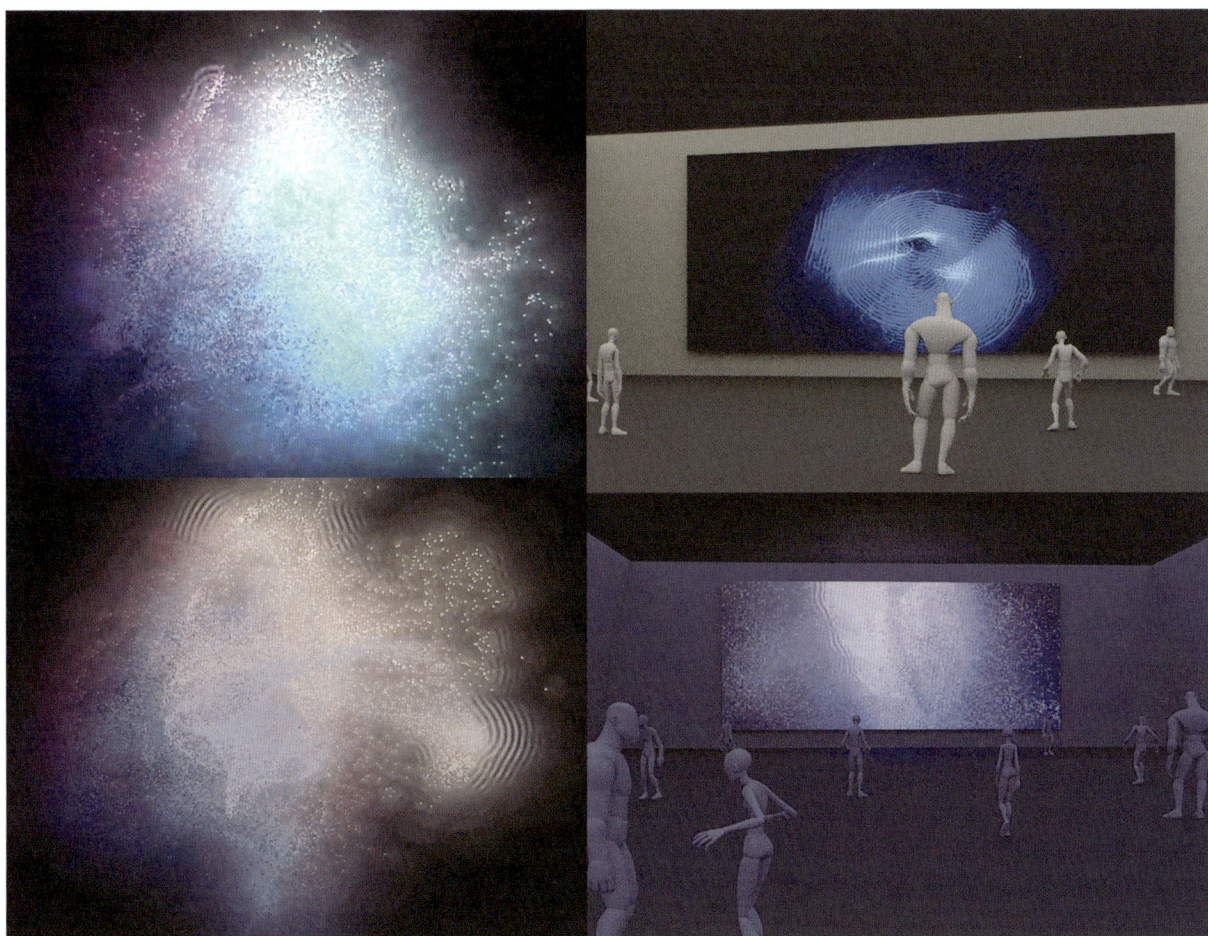

2.2.15.2　城语

2.2.15.2　The Sounds of The City

作者：杜昊峪、白易诺、王晓楠

By Haoyu Du, Yinuo Bai, Xiaonan Wang

　　"城语"的主题是以未来主义为背景的城市噪声可视化表达。创作的出发点是人和环境关系的观照——通过回溯城市发展的历史，从而反思在城市扩张中人与日常环境之间的脱节，从噪声角度重构人与城市环境的有机联系。在项目制作中，我们主要采样了不同时空的城市声音，例如20世纪80年代的叫卖声、自行车铃声和现代的飞机起飞音，汽车鸣笛声等，并以此为基础重新编曲与创作，利用数字媒体和实拍影像，完成了一部视听一体化的作品。

　　The theme of our project, "The Sounds of The City" is the visual expression of urban noise against the background of futurism. The starting point of creation is based on the care of the relationship between people and the environment. By tracing back the history of urban development, we can reflect on the disconnection between people and the daily environment in urban expansion and reconstruct the organic relationship between people and the urban environment from the perspective of noise. In the project production, we mainly sampled urban sounds of different times and spaces, such as the peddling voice in the 1980s, the ring of bicycles, the takeoff sound of modern airplanes, and the whistle sound of cars. Based on this, we composed and created a work of audio-visual integration using digital media and live video.

2.2.15.3　花墙乐队品牌形象设计

2.2.15.3　Wall Band Brand Design

作者：林偕郁、常臣晨、王心睿、李敏源

By Xieyu Lin, Chenchen Chang, Xinrui Wang, Minyuan Li

　　根据对乐队的音乐风格以及演出方式进行调研分析，提炼出风格关键词：新潮有趣，强烈视觉性，叛逆大胆以及年轻热血。根据这些关键词进行视觉设计，受众人群为喜爱潮流音乐及音乐演出现场体验的年轻消费者。在整体视觉中，手绘线稿体现新潮有趣，抽象图形带来强烈的视觉性，手写体体现乐队音乐创作的叛逆大胆，高饱和度的色彩彰显该乐队音乐的年轻热血风格。视觉产出内容有：标语，门票，宣传单，海报，名片，专辑封面，横幅等。

　　Based on the investigation and analysis of the band's music style and performance mode, the style keywords are extracted: trendy and interesting, strongly visual, rebellious, bold, as well as young and enthusiastic. Visual design is carried out according to these keywords, and the audience is young consumers who love trendy music and live house experience. In the overall vision, the hand-painted live house reflects the trendy and interesting, the abstract graphics bring a strong sense of vision, the handwritten font reflects the rebellious and bold music creation of the band, and the high saturation color highlights the young and enthusiastic style of the band's music. The visual output contents include logos, tickets, leaflets, posters, business cards, album covers, banners, etc.

2.2.15.4　困在时间里的尘埃

2.2.15.4　Stranded

作者：王家轩、王佳慧、韩薇、栾承骏、黄雨欣、曹雯

By Jiaxuan Wang, Jiahui Wang, Wei Han, Chengjun Luan, Yuxin Huang, Wen Cao

　　阿尔茨海默病（以下简称AD）是常见于老年人的神经退行性疾病，但大多数人对阿尔兹海默病的了解并不深刻，对其敬而远之。AD患者的记忆就像尘埃一样缥缈不定，因而我们的作品将患者的记忆拟为一粒尘埃。通过音乐可视化来表现AD患者在病发时每个阶段的困境，希望来唤起人们对AD患者的同理心。

　　想象中的AD就像一架发条松了的八音盒，随着旋转小人慢慢竭力地转动，但依然敌不过物理上的衰败，最后结尾回到坏掉的八音盒音色，暗示患者患病的状态。希望能呼吁大家重视AD，及早预防、积极治疗。这是我们对于这段音乐的构想。

　　Alzheimer's disease, short for AD, is a common neurodegenerative disease in the elderly. Still, most people do not have a deep understanding of Alzheimer's disease and stay away from it. The memory of patients with AD is as uncertain as dust, and our work describes the memory of patients as a grain of dust. Music visualization is used to show the plight of patients with AD at each stage, hoping to arouse people's empathy for them.

　　When I think of AD, I picture it as an eight-tone music box with loose clockwork. Although the figurine keeps revolving, it can't compete with the physical decline. Finally, I returned to the broken 8-tone music box, to imply that the patient was sick. Through this, I hope we can raise awareness and direct everyone's attention to AD, to promote early detection, prevention, and active treatment. This is my reasoning behind this piece of music.

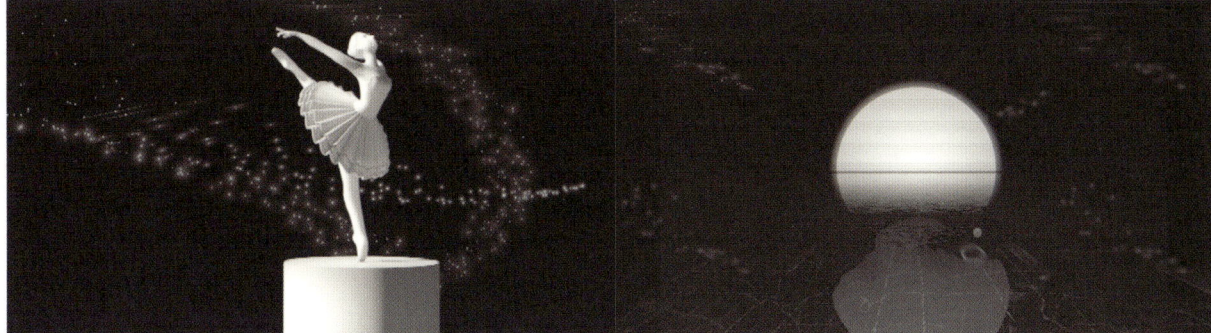

↘ 2.2.16 课题：购买本土产品——图形标识

2.2.16 TASK: BUY LOCAL—GRAPHIC IDENTITY

课题说明：

设计一种标识和材料以支持邻里之间的经济往来，并帮助建立一个以本地思维为主的社区，从而促进社区的繁荣发展。学生需要能解读自己所在社区的身份价值观，并将其转化为图形和可持续包装理念，以创造情感共鸣，吸引目标受众。管理时间、材料和资源，表现出解读文化倾向的能力，并运用沟通能力来完善最终成果。我们建议学生考虑丝网印刷、凸版胶印、织物染色等，将其作为与数字化生产结合使用的工具。这些需要亲自实践的工作使参与者能够深入了解设计的敏感性，并且也使参与者能够对制作流程了如指掌。其他方面的内容包括对设计制作实用性相关问题的思考，以及通过工作坊实践、评判和讨论来探索色彩理论及其应用。除了概念性的工作，我们还将介绍一些在设计制作中必要的技术，以及如何与数字设计相结合的方法。

指导老师：

伊曼纽尔·巴博萨，葡萄牙马托西纽什艺术与设计学院副教授和研究员

Task Description:

Design an identity and materials to support an economy of friends and neighbors and help to build a community that thrives by thinking local first. Students must be able to interpret the identity values of their local regions, adapting them to graphics and sustainable packaging ideas to create emotional responses and seduce target audiences. Managing time, materials, and resources, showing the ability to interpret cultural tendencies, and improving the communication ability capacity to improve results. Students are suggested to consider screen printing, letters, fabric dyeing, etc. as tools that can be used in conjunction with digital production. They are hands-on ways of working that will give insights into design sensibilities and enable you to be fully in charge of how things are made. Other aspects include considering issues surrounding design production's usefulness and exploring color theory and its applications through studio practice, critiques, and discussion. As well as this conceptual work, we will cover necessary technical aspects of design production and how these interface with digital design.

Tutor:

Emanuel Barbosa, Associate Professor and Researcher, ESAD MATOSINHOS, Portugal

小店再生计划

Little Shops Remade

作者：陈易知、林鸿峰、Byeungho Choi、吴颖晗、邹雅婧

By Yizhi Chen, Hongfeng Lin, Byeungho Choi, Yinghan Wu, Yajing Zou

　　我们会按照顾客逛集市的逻辑顺序对我们的设计进行展示。首先是街道宣传，人们会在公交车站和路边看到这样的海报，使人们初步感受到集市的丰富内容。接下来，当人们抵达集市的时候，他们会看见颜色缤纷的门头设计，吸引人们进入集市。进入集市的街道时，他们会看到来自不同文化所带来的丰富内容。人们在集市中参与活动，会获得一些活动周边礼物。最后是集市中的活动。我们设计了一个游戏来帮助人们更好地探索社区中的小店，并借助手册展示游戏的玩法；游戏道具设计内容包括地图和任务卡。

　　We will show our design according to the logical order of customers visiting the market. Posters like these can be seen at bus stops and on the roadside, giving people a taste of the market's rich content. Then when people arrive at the market, they will see the colorful gate design to attract people to the market. As they enter the street of the market, they see the richness brought by different cultures. People participate in the activities in the market and get some gifts around the activities. Finally, the activities in the market. We have designed a game to help people better explore small stores in the community. First, there is a manual showing how to play the game, and then there is our game props design, including maps and task cards.

↘ **2.2.17　课题：设计师宣言：内容即信息**

2.2.17　TASK: DESIGNERS MANIFESTO: THE CONTENT IS THE MESSAGE

课题说明：

　　该工作坊旨在利用赫伯特·拜尔（Herbert Bayer）首创的心、脑、眼协同的传统设计思维工具来加强参与者的序列化思维能力，它由蒂姆·布朗（Tim Brown）和 *Inside The Box* 的作者雅各布·戈登伯格（Jacob Goldenberg）重新包装，并以"设计思维"的形式呈现。该工作坊的主要目的是通过深入研究来打破人们固有的思维、情感和意愿方式。你的宣言是什么？你有个人宣言想要分享吗？你一直敬仰的宣言是什么样的？

指导老师：

　　Lefteris Heretakis，英国新白金汉郡大学平面设计与插图系主任

Task Description:

　　This workshop is designed to strengthen the sequential thinking ability of the participants using the traditional tools of design thinking of Heart, Heart, and Eye coordination as pioneered by Herbert Bayer, repackaged by Tim Brown and Jacob Goldenberg of *Inside the Box*, and presented as "Design thinking". The primary aim of this workshop is to unblock established ways of thinking, feeling, and willing through a deep examination of their processes. What is Your Manifesto? Do you have a personal manifesto that you'd like to share? How about a manifesto that you've always admired?

Tutor:

　　Lefteris Heretakis, Dean of Graphic Design and Illustration at Buckinghamshire New University, UK

不完全设计师成长手册

Incomplete Designer Growth Manual

作者：胡培坤、顾润红、李婧煊、Athhar Izza

By Peishen Hu, Runhong Gu, Jingxuan Li, Athhar Izza

　　我们在本次课题中提出了一个概念，叫：不完全设计师成长手册。设计师宣言是设计师在成长过程中凝练所得，并不断更新迭代，由此对应"不完全"与"成长"两个词。设计师在成长过程中同样也会遇到难题，这个时候我们希望能给他们一些帮助。我们最终选择了软件这一载体来承载我们心中的设计师宣言。希望使设计师们能在此感受到共鸣，从而形成一个开放性的设计师答案之书。设计是为构建有意义的秩序而付出的有意识的直觉上的努力，设计师应当将信息价值转变为内容价值，包括从众多的日常琐碎信息中筛选出有用的信息，深入了解和学习。设计者们深知生活是设计的基础；设计者们在生活中寻找设计的源泉与素材。本次课题想从生活本身出发，发掘其中蕴藏的可能性，并结合六句宣言，用视觉语言呈现出来。

　　本次项目将延续四个不同的思路：数字媒体艺术设计、时尚设计、视觉传达设计、室内设计。我们将运用各自的专业的视觉语言，结合教授教给我们的方法论来构建自身的宣言，最后形成一个合集。我们希望这个合集是有针对性的，能够让不同专业方向的设计师找寻对应自己专业宣言的一个区域。目前只有四个区域，未来可能随着更多设计师的加入，再添加更多的区域，从而形成一个开放性的设计师"答案之书"。

　　In this task, we express the manifestos by combining the design thinking and methods taught by the professor. Therefore, we put forward a concept called The Incomplete Designer Growth Manual. The designer manifestos are a concise result of the designer's growth process, and it is constantly updated and iterated, thus corresponding to the two words "incomplete" and "growth." Designers will also encounter difficulties in their growth, and we hope to have something to help them. By that, we decided to choose software as a medium to carry designer's manifestos in our hearts. We hope that with this open and inclusive book of designer's answers, many designers can feel the resonance here. Design is a conscious and intuitive effort to build meaningful order, and designers should transform the value of information into the value of content. Designers should transform the value of information into the value of content, sifting through everyday trivial information to find useful information and going deeper to understand and learn. Designers know that life is the basis of design; designers look for sources and materials in life. So, we wanted to start with life itself, discover its possibilities, combine them with my six statements, and present them in a visual language.

　　This project will continue along four lines of inquiry: Digital Media Art, Fashion Design, Visual Communication Design, and Interior Design. We will employ our respective professional visual languages, combined with the methodologies taught to us by our instructors, to construct our manifestos. Ultimately, a collection will be formed, and this is because we hope the collection will be targeted, enabling designers from different specializations to find a section that corresponds to their own professional manifesto. Currently, there are only four sections, but in the future, as more designers join, additional sections may be added, thus creating an open "book of answers" for designers.

↘ **2.2.18 课题：设计向善**

2.2.18 TASK: DESIGN FOR GOOD

课题说明：

在如今这个不断变化的社会中，我们面临着许多影响我们个人和社会的问题。因此，本工作坊邀请你考虑视觉传达设计如何在这些问题上有所作为。提出的问题包括：作为平面设计师，我们是否能够为世界带来改变？我们的设计可以在面临的哪些领域和挑战中产生积极的影响？作为一个设计师，什么对你和你所生活的社会是重要的。从技术、可持续性、政治或文化的角度考虑这些问题。反思你作为一个设计师的角色，并探索你的创意实践如何能与好的设计联系起来。你将被要求考虑当前对你很重要的社会、文化问题。这可能是非常个人化的东西，也可能是更大的全球规模。这个研讨会要求你探索这些问题可能是什么，并进行一系列的研究。

指导老师：

丽莎·温斯坦利，新加坡南洋理工大学艺术设计与媒体学院助理教授

Task Description:

In today's ever-changing society, we face many issues that affect us individually and on a societal basis. Therefore, this workshop invites you to consider how visual communication design can contribute to these issues. Raising questions such as: Can we, as graphic designers, make a difference in the world? What areas and challenges do we face where design can positively impact? What is important to you as a designer and the society you live in? Consider these questions from technological, sustainability, political, or cultural perspectives. Reflect on your role as a designer and explore how your creative practice can connect to design for good. You'll be asked to consider the current societal or cultural issues that are important to you. This could be something very personal or on a larger global scale. This workshop asks you to explore what those issues might be and to conduct a body of research.

Tutor:

Lisa Winstanley, Assistant Professor, School of Art, Design and Media, Nanyang Technological University, Singapore

爱与讽刺

The Erotic Irony

作者：Monica Suryanto、汤苓、崔馨予、Zefa Christiansa Patricia Malau、Yeon Joo Lim

By Monica Suryanto, Ling Tang, Xinyu Cui, Zefa Christiansa Patricia Malau, Yeon Joo Lim

性骚扰经常出现在工作场合和大学里，但这种伤害也会发生在童年时期。女性从小遭遇的性骚扰将间接地给她们的思维带来深刻的烙印，直到成年后。当然，恐惧、焦虑和创伤会一直困扰着她们，让她们感到愧疚。我们想依托本次工作坊邀请读者和艺术鉴赏者与公众进行话题分享。这个问题可能总是被一些人认为是小事，甚至认为不存在，但这可能是一个人的核心记忆。我们必须时刻注意周围的环境，敢于发声。我们要勇于责备、指责，并告诉对方性骚扰是错误的。

Sexual harassment often occurs in the workplace and at univer sities, but the harm can also occurs in childhood. The source of sexual harassment that female receive from childhood will indirectly give a deep stigma to her thinking until adulthood. Of course, fear, anxiety, and trauma will always haunt them with feelings of guilt. With this work, we invite readers and art connoisseurs to share this with the public. This issue may always be considered minor by some or even regardede as nonexistent. However this can be a person's core memory that they take into consideration when making decisions in their lives. We must always be aware of our surroundings and dare to voice our valuable voice. We need to start having the courage to reprove, accuse, and tell the person that sexual harassment is wrong.

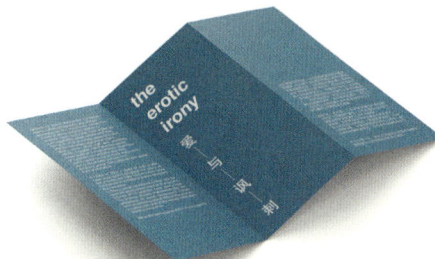

↘ **2.2.19 课题：数字时代的数字时尚与数字传播**

2.2.19 TASK: DIGITAL FASHION AND COMMUNICATION IN DIGITAL AGE

课题说明：

数字时装是利用计算机技术和3D软件构建的服装的可视化表示。由于道德意识的提高以及人工智能等数字时尚技术的使用，该行业正在崛起。数字时尚也是数字技术和时装之间的相互作用。信息和通信技术（ICT）已经深刻地融入了时尚行业以及客户和前景的实践。时尚行业总体上为数字时尚的引入铺平了道路，更多的技术正在这个行业中出现，比如虚拟化妆间和时装业的游戏化。围绕元宇宙和时尚的对话是多层次的。通过数字时尚，我们可以体验最纯粹的时尚，没有功能障碍。

Task Description:

Digital Fashion is the visual representation of clothing built using computer technologies and 3D software. This industry is rising due to ethical awareness and the use of digital fashion technology, such as artificial intelligence to create products with complex social and technical software. Digital fashion is also the interplay between digital technology and couture. Information and communications technology (ICT) have been deeply integrated into the fashion industry and within the experience of clients and prospects. The fashion industry, in general, has paved the way for digital fashion to be introduced with more technology being in the industry, like virtual dressing rooms and the gamification of the fashion industry. The conversations around the metaverse and fashion are multilayered. With digital fashion, we can experience its purest form, devoid of functional barriers.

指导老师：

朴志善，韩国祥明大学公共设计中心特聘教授
于晓洋，北京服装学院动画专业教师

Tutor:

Jisun Park, Distinguished Professor of Public DesignCenter, Sangmyung University, Republic of Korea

Xiaoyang Yu, Lecturer of Beijing Institute of Fashion Technology

WWW

WWW

作者：董书豪、蔡安东、许珮慈、Sunghye Ahn、Doyun Kim、Chaewon Shin

By Shuhao Dong, Anthony Tsai, Patsy Hsu, Sunghye Ahn, Doyun Kim, Chaewon Shin

以"集体智慧与万物永生"概念为中心的元宇宙虚拟世界。这是一个重新定义数字时代下的数字时尚与数字传播的新型虚拟时尚设计，由韩国祥明大学的朴智瑄教授与北京服装学院于晓洋老师指导，中韩两国学生协同合作的设计课题。课题围绕现代人类的"精神""欲望"以及"新型通信方式"，结合生态学和自然视觉资源实现一个和合共生的数字新生态世界，名为"WWW"。

A meta-cosmic virtual world centered on "collective intelligence and eternal life". This is a new virtual fashion design that redefines digital fashion and digital communication in the digital era. It is a collaborative design task between Chinese and Korean students under the guidance of Professor Park Jisun from Sangmyung University and Professor Yu Xiaoyang from Beijing Institute of Fashion Technology. The task focuses on the "spirit" "desire" and "new communication mode" of modern human beings, combined with ecology and natural visual resources to achieve a new digital ecological world relying on human wisdom to achieve harmony and symbiosis, named "WWW".

↘ **2.2.20 课题：通过九型人格特征进行自我发现**

2.2.20 TASK: SELF DISCOVERY THROUGH ENNEAGRAM BRANDING

课题说明：

　　九型人格已经成为自我发现和理解的有力工具。它概述了九种不同的人格类型，描述了一个人如何在核心恐惧和欲望的基础上与自我、他人和世界产生链接。通过这个项目，你将确定你的九型人格类型，并开发一个迷你身份系统，唤起你人格类型的一个或多个特征。在这个项目中，我们使用九型人格作为基础理解的形式，区分特征并探索如何在视觉上表达它们（无论是字面的还是概念的）是身份设计的核心。九型人格在当今顶级科技 CEO 中的普及反映了它作为领导者和团队的强大工具的价值。归根结底，当我们更好地了解自己和我们的员工时，我们就能以更有效的方式进行领导。通过了解彼此的九型人格类型，我们就有了开启核心动机的钥匙，从而加强工作关系，推动实现共同的成功愿景。

Task Description:

　　The Enneagram has emerged as a powerful tool for self-discovery and understanding. It outlines nine distinct personality types that describe how one relates to self, others, and the world based on core fears and desires. With this project, you will identify your enneagram and wing types and develop a mini-identity system that evokes one or more characteristics of your type. We use the Enneagrams as a form of base understanding for this project but distinguishing characteristics and exploring how to express them visually (literally or conceptually) is the heart of identity design. Enneagram's popularity amongst today's tech CEOs reflects its value as a powerful tool for leaders and teams. Ultimately, we can lead more effectively when we understand ourselves and our people better. By understanding one another's Enneagram type, we have the keys to unlocking our core motivators, strengthening working relationships, and driving toward a shared vision of success.

指导老师：

　　瑞安·斯洛恩，美国阿肯色大学艺术学院平面设计系助理教授

Tutor:

　　Ryan Slone, Assistant Professor of Graphic Design, College of Art, University of Arkansas, USA

人格实验室

Personality Studio

作者：刘欣雪、董潇雨、高沛琪、石弋萱、Viona Nadya Karenina

By Xinxue Liu, Xiaoyu Dong, Peiqi Gao, Yixuan Shi, Viona Nadya Karenina

　　九型人格测试有助于人们更好地认识自我、了解他人。我们团队利用自身的"人格"元素进行字体与标识的可视化的表达设计：1w9是理性、自控，在字母"E"和"F"之间做出连接的变化，改变单词的排版，从而表达严肃却又富含设计的风格。3w2是自信、魅力，以切分感和线条感的字体，表达果断和远见。3w4是高效、完美主义，将罗马柱与电路板作为视觉符号，表达干练与完美平衡。5w4是反叛、独立，为其设计的标志突出"奇怪"的特征，具有神秘感，符合大众对"学者"的印象。6w7是忠诚、纠结，采用了纠缠与粘连的视觉效果，体现人格的纠结与依附。项目最终以此为媒介把这种个性的图案字体运用在不同品牌的设计之中，赋予品牌人格。

　　Taking the "Enneagram test" helps people better understand themselves and others. Our team uses its own "personality" elements to carry out visual expression design of fonts and logos: 1w9 is rational and controlled, making a connection between the letter's "E" and "F", changing the typography of words to express a severe but design-rich style. 3w2 is a confident, charismatic, syncopated, linear font expressing decisiveness and vision. 3w4 is efficient and perfectionist, using Roman Columns and circuit boards as visual symbols to express competence and perfect balance. 5w4 is rebellious and independent. The logo designed for it highlights the characteristics of "strange," has a sense of mystery, and conforms to the public's impression of "scholars." 6w7 is loyalty and entanglement, using the visual effects of entanglement and adhesion, reflecting the entanglement and attachment of personality. Finally, the project uses this personalized pattern and font in different brand designs and endows the brand with personality.

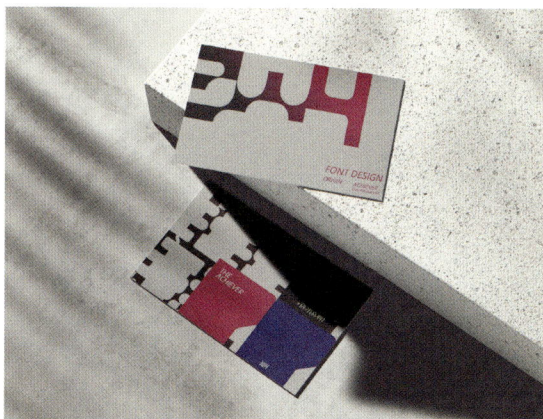

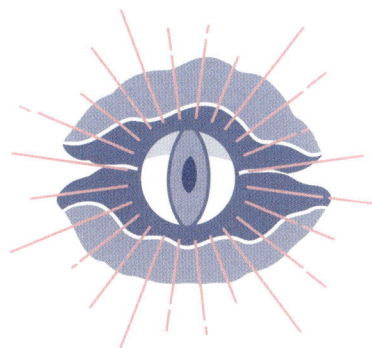

- OBSERVER -
观察者

↘ 2.2.21 课题：通过无声之物：重塑奥地利木偶表演

2.2.21 TASK: THROUGH THE SILENT OBJECTS: RE-DISCOVER THE MANIFESTATION OF AUSTRIAN PUPPETRY

课题说明：

木偶戏是一种戏剧或表演的形式，它涉及对木偶的操纵。木偶由木偶演员制作或操控。Teschner 的创作是一个明显的例子，两个木偶的表演形式和技术可以在设计环境中形成全新的混合方式，这体现了奥地利和印度尼西亚两国的文化和审美观点。学生们将从视觉、叙事、历史、哲学等多个方面探索并加深他们对木偶戏的理解。他们还可以从学科角度探索，如木偶的美学观点（服装、时尚和材料），或者从剧场舞台的空间设计角度来探讨。学生还可以思考音乐或音响、演奏和动作等许多方面。本项目希望鼓励他们充分发挥自己的感官潜能。在小组合作的同时，学生们将能够在当代背景下表达他们的集体思想，并将其应用到最终作品中。最终成果将是多样的，这代表着参与者的再发现和创新表现。

指导老师：

汉尼·维加亚，奥地利林茨艺术大学副教授

阿玛利亚·巴博萨，奥地利林茨艺术大学教授

Ákos Hutter，匈牙利佩奇大学建筑学院工程和信息技术学院教授

Task Description:

Puppetry is a form of theatre or performance that involves manipulating puppets animated or manipulated by humans called puppeteers. Teschner's creation is a clear example that both puppets' mixture of forms and techniques can exhibit a new hybrid in design contexts that manifest the cultural and aesthetic perspectives of both Austria and Indonesia. Students will explore and discover their interpretation of puppetry from various aspects, such as visual, narrative, history, philosophy, etc. They also can explore from their disciplinary perspective, like the aesthetic view of the puppets (costumes, fashion, and materials) or from the spatial design perspective of the theatre stages. Students can also think about the music or sound, play and movement, and many other aspects. This project will encourage them to play with their senses to their full potential. While working together in their group, students can manifest their collective thoughts in the contemporary context and apply them to their final artwork. The outcomes will be diverse, representing the participants' re-discovery and new manifestation.

Tutor:

Hanny Wijaya, Associate Professor, University of the Arts Linz, Austria

Amalia Barboza, Professor, University of the Arts Linz, Austria

Ákos Hutter, Professor, Faculty of Engineering and Information Technology, Faculty of Architecture, University of Pécs, Hungary

重塑奥地利木偶表演

Re-Discover the Manifestation of Austrian Puppetry

作者：Abel Alvieto、Angeline Yuliani、Laura Valencia Kamara、尹涵、张慧丰、Mohammad Reza Fahlevi Nugraha

By Abel Alvieto, Angeline Yuliani, Laura Valencia Kamara, Han Yin, Huifeng Zhang, Mohammad Reza Fahlevi Nugraha

当今时代，木偶戏正逐渐被人们淡忘。人们社会互动的频率有所下降，这促使我们进行创新，加强互动，促成优雅的生活。基于这两点，我们创造了 PUPPET-TRY。PUPPET-TRY 是一个户外公共装置，旨在基于当地文化的视觉效果以数字化图形的形式呈现一种全新的木偶表演方式，我们还重新诠释了"Kasperle"木偶的本质，它具有娱乐性、可爱性和灵活的特点，这是 PUPPET-TRY 作为一个互动装置如何被用户感知的基本理念。因此这个公共装置是可以连接人们（从儿童、年轻人到老年人），帮助他们拥抱幸福，并创造出可持续环境的。我们想要带来的可持续是指改善人们的健康，并使用"运动"作为装置的能源。"WE ARE PUPPETY"意味着自我即木偶。人随意做出动作，它将在 LCD 屏上以木偶的形式出现。这将联系起不同的人，并最终将不同人所做的动作进行组织编排。

The number of authentic puppet shows is slowly getting dimmer in this modern era. Furthermore, the frequency of social interactions that has decreased since the pandemic motivates us to make an innovation that can make people interact in this post-pandemic era and increase their graceful life. Based on these two main points, we made PUPPET-TRY. PUPPET-TRY is an outdoor public installation that presents a new way of performing puppets in the form of digitizing graphics based on local cultural visuals. We also reinterpret the essence of the "Kasperle" puppet which has an entertaining, lovable, and flexible persona for the basic philosophy of how users can perceive PUPPET-TRY as an interactive installation. This public installation connects people (from the children, younger, to the older generation), improving their well-being and creating a sustainable environment. We want to bring a sustainable aspect to improve people's health and use a "movement" as this installation energy. "We are the puppets" means that the user will be the puppet. The user will move and appear on the LCD screen as a puppet. This goes along with another user until they make a group choreography.

↘ **2.2.22 课题：优雅生活烹饪书**

2.2.22 TASK: GRACEFUL LIFE COOKBOOK

课题说明：

设计一本不仅具有视觉冲击力，而且能展示您创造力的书籍。你是这本书的作者。它包含了食谱和视觉效果，对你来说它是如此有趣以至于你无法停止发挥、设计它。它可以成为你的作品集的重要补充，但它首先是一个个人项目；是你有兴趣创造和设计的东西——它必须反映出烹饪食物、饮食和生活的优雅方式。设计一本书涉及对细节的高度关注——你需要知道你的读者会被什么东西所吸引，并花时间和精力把它作为一件艺术品来创作。视觉效果是非常重要的。你可以使用任何形式，从古老的植物学插图到社论摄影、印刷品甚至拼贴画。记住你是在设计一本食谱，但不要忘了食物本身也有历史和文化，所以如果你认为你的书可以做一些独特的页面或传播，就将其可视化，不要害怕尝试。

指导老师：

Phil Cleaver，英国密德萨斯大学创意产业教授

Task Description:

Designing a book that is not only visually stunning but showcases your creativity. You are the author of this book; it contains recipes and visuals, and it wants to be so interesting to you that you can't stop playing/designing it. It can be a valuable addition to your portfolio, but it is, above all, a personal project that you are interested in creating and designing–and it must reflect a graceful way of cooking food, eating, and living. Designing a book involves acute attention to detail – You need to know what your audience will be visually captured by and put the time and effort into creating it as a piece of art. Visuals are very important. You can use anything from botanical illustrations to editorial photography, prints, or even collage. Remember that you're designing a cookbook, but don't forget that food has history and culture within its right, so if you think your book could do with the odd page or spread that visualizes this, don't be afraid to experiment.

Tutor:

Phil Cleaver, Professor of Creative Industries, Middlesex University, UK

优雅的烹饪生活书

Graceful Life Cookbook

作者：刘梦竹、Maharani Puspita Ayu Gunhadi、洪嘉祺、朱冠宇

By Mengzhu Liu, Maharani Puspita Ayu Gunhadi, Jiaqi Hong, Guanyu Zhu

我们的设计灵感来自食品塑料污染，这是全球最严重的环境问题之一。我们认为当下我们需要的食谱是一本有环保意识的食谱，这样我们就能拥有更优雅的低碳饮食。我们设计了一份带着二维码的食谱，通过扫码可以得到一个食物盒子。盒子中包含食谱所需的食材与佐料，如果您继续购买并收集食谱卡，每周还可获得四份推荐食谱。我们选择能减少碳足迹的饮食方式，比如用电动车送食材，用可食用包装，使用当地食材等——这就是我们的"优雅的烹饪生活书"。

Our design inspiration comes from disposable plastic pollution, one of the most severe environmental problems in the world. The recipe we need now is an environmentally conscious cookbook to have a more graceful low-carbon diet. We designed a recipe with a QR code; you can scan it for a food box. It contains the ingredients and seasonings needed for the recipe and four recommended recipes per week if you continue to buy and collect the recipe cards. We choose to eat in ways that reduce our carbon footprint, such as using electric cars to deliver food, edible packaging, local ingredients, etc. This is our Graceful Life CookBook.

디자인 컨셉
设计理念
Design Concept

Come order the edible meal boxes!

快来订购这款可食用的餐盒！

이 먹을 수있는 도시락을 주문하십시오!

↘ 2.2.23 课题：有意义的设计：在虚拟世界中设计的"优雅生活"的构建

2.2.23 TASK: MEANINGFUL DESIGN: THE CONSTRUCTION OF A GRACEFUL LIFE DESIGNED IN A VIRTUAL WORLD

课题说明：

随着沉浸式体验得到更广泛的应用，必须考虑如何利用它们来增强人类的积极体验。VR和AR已经被广泛用于增强一系列的人类体验，包括培训、医疗和更传统的娱乐。设计过程和方法也变得更加细微，使用更清晰的策略和方法来设计这种体验，从而挖掘出人类的情感。在人类经验的中心，记忆和怀旧丰富了人类的功能，情感在我们的交流系统中起着核心作用。具体而言，在"优雅地生活"这个总体主题中，它挑战了沉浸式环境中的情感设计如何能像对非数字空间中的人工制品的反应一样，衍生和唤起情感。在这个项目中，学生们将考虑他们的设计过程和方法，以推动数字体验的发展，从而获得一个与情感相关的概念，如怀旧。学生将需要考虑不同层次的设计方法，这些方法与创造一个沉浸式的体验相交织。

Task Description:

As immersive experiences become more widely used, it is imperative to consider how they can be used to augment positive human experiences. VR and AR have been widely used to augment a range of human experiences from training, medicine, and traditional entertainment. The design process and approaches have become more nuanced with clearer strategies and approaches to designing such experiences that tap into human emotion. Central to the human experience, memory and nostalgia enrich human function with emotions playing a central role in our communication system. Specifically, this overarching theme of a Graceful Life, challenges how emotional design in immersive environments can derive and evoke emotions like responses to artifacts in a non-digital space. In this project, students will consider their design process and approach in the context of pushing the envelope of a digital experience designed to illicit an emotionally linked concept such as nostalgia. Students will need to consider different layers of design approaches that weave into creating an immersive experience.

指导老师：

Jonathan Pillai，资深学者，澳大利亚科廷大学动画和游戏课程讲师

Anne Farren，澳大利亚科廷大学设计学院教授

Stephanie Hampson，澳大利亚科廷大学设计与建筑环境学院讲师

Tutor:

Jonathan Pillai, Senior Scholar, Lecturer, Animation and Games at Curtin University, Australia

Anne Farren, Professor, School of Design, Curtin University, Australia

Stephanie Hampson, Lecturer, School of Design and the Built Environment, Curtin University, Australia

蝴蝶效应

The Butterfly Effect

作者：Sundari Kayani Candijaya、罗霄雯、黄柏敏、Damon Drage、Vivien Tanuwidjaja、Xuan Ying Lee

By Sundari Kayani Candijaya, Xiaowen Luo, Baimin Huang, Damon Drage, Vivien Tanuwidjaja, Xuan Ying Lee

　　"蝴蝶效应"的主要概念是将用户带入不同的世界，通过使用谜题，探索不同的区域，与过去不同的怀旧片段互动。怀旧的感觉可以让人在当下感觉更好，并且有可能让他们以乐观的态度面向未来。通过反思曾经的事情，许多人可以从精神低谷中恢复过来，获得面对未来的勇气。玩家将被传送到的世界是海滩和美食仙境。蝴蝶将被用作一种运输手段，通过触摸它们进入新的世界。在这些环境中，将有互动的物体和谜题需要解决，以取得下一步进展。我们的团队希望给用户留下一个结局印象。假设有人确实需要经历怀旧才能找到乐观的未来。在这种情况下，这一新的积极源泉不应被用来助长压力，使其成为不可能达到的标准。相反，我们的最终目标是提醒我们的用户，接受你现在的生活状态才是你需要的，你过去的成就应该得到赞美。

　　The main concept of "The Butterfly Effect" is to transport the user into different worlds, using puzzles, and explore the different areas that interact with different nostalgic pieces of the past. The feeling of nostalgia can make one feel better now and has the potential for them to be more optimistic about the future. By reflecting on what once was, many can recover from their mental slump and gain the courage to face the future. The players will be transported to the Beach and Food Wonderland worlds. Butterflies will be used to transport to new worlds by touching them. In these environments, there will be interactive objects and puzzles to be solved to progress further. Our team wanted to leave an ending impression on our users. Suppose someone does need to experience nostalgia to be able to find optimism in the future. In that case, this new source of positivity should not be used to fuel the pressure of becoming the impossible standard of yourself. Instead, our final goal is to remind our users that accepting where you are in life is where you need to be, and your past accomplishments should be celebrated.

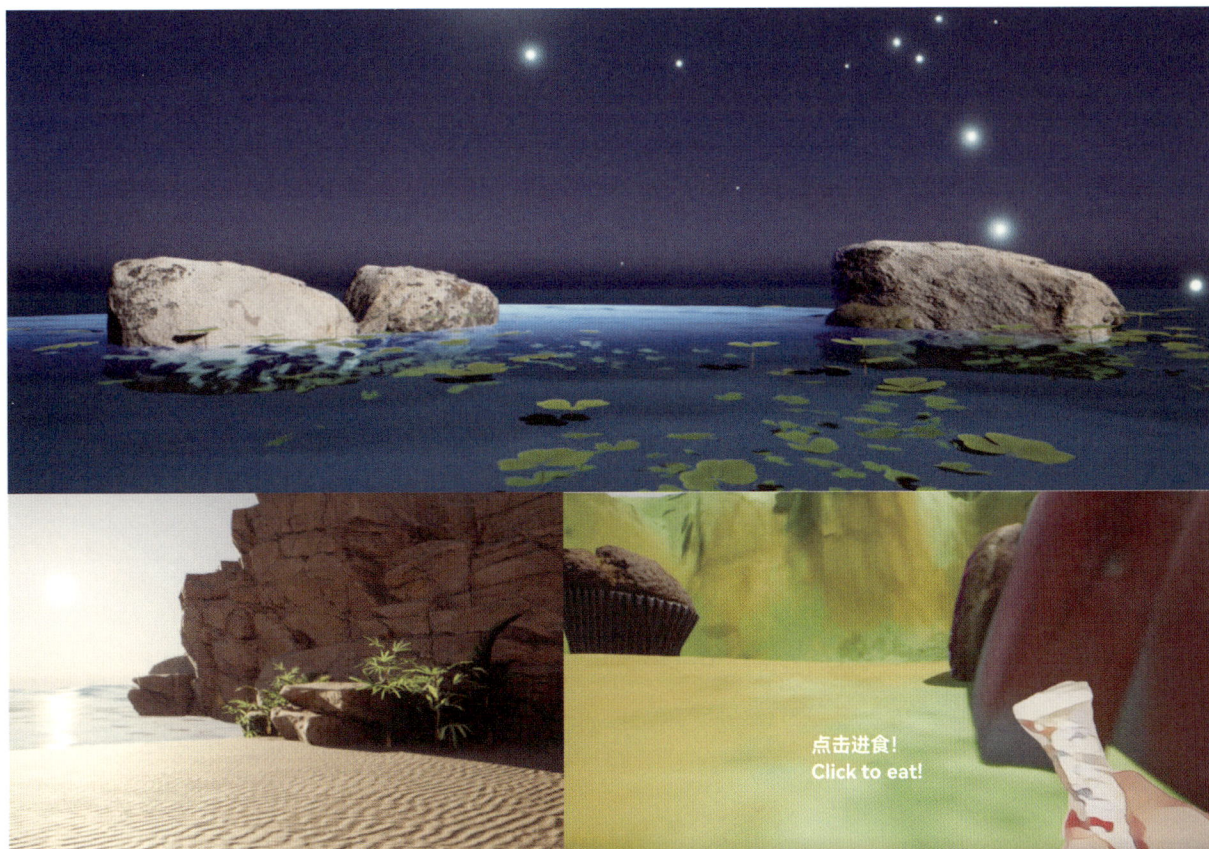

点击进食！
Click to eat!

03

2021 设计马拉松
2021 DESIGN DAY MARATHON

3.1 2021 设计马拉松总体安排

3.1 2021 DESIGN DAY MARATHON GENERAL ARRANGEMENTS

3.1.1   主题：设计可持续的福祉

2021年的设计马拉松主题为"设计可持续的福祉"。人类的福祉（Human well-being）指的是健康幸福和物质上富足的生活状态。当前，我们常关注的可持续发展目标，其最终目的是提高人类福祉，满足当代人和后代人在物质和精神这两方面的需求。随着物质水平的提高，我们的社会出现了高度追求健康生活与幸福感的福祉族（Well-beinger）群体。福祉族追求的不仅是物质上的满足，还希望通过调节心灵和身体来达到更为健康与幸福的生活状态。

3.1.1   THEME: DESIGNING FOR SUSTAINABLE WELL-BEING

The 2021 Design Day Marathon theme is "Designing for Sustainable Well-Being." Human well-being refers to a living state of health, happiness, and material abundance. The sustainable development goals, which we often focus on today, are ultimately aimed at improving human well-being and meeting the material and spiritual needs of present and future generations. With the improvement of material level, there are well-being groups in our society who are highly pursuing healthy life and happiness. The Well-beinger pursues material satisfaction and a healthier and happier life by adjusting the mind and body.

3.1.2 2021 设计马拉松的构成

北京服装学院
科学艺术
时尚周

协会、学会

企业赞助方

高校

媒体

企业

同学

教师

独立设计师

北京国际设计周

国内外高校

机构单位

组织

参与

前沿

落地

包容

设计马拉松

时尚

实验

高效

主题

2017

2018

2019

2020

2021

24小时快速设计

设计更好的银发
互联网用户体验

青银共创未来

设计马拉松
即兴直播间

设计可持续福祉

品牌设计
影视设计
空间设计
产品设计

网站使用体验
社交软件体验
智能设备体验

新闻设计
食物设计
老龄设计
影视设计
空间设计
产品设计
服务设计
未来设计
创新适老设计
生活方式设计

创新创业
空间设计
服务设计
食物设计
艺术创作
健康设计
老龄设计
设计方法
圆桌会议
设计领导力
可持续设计
可持续时尚

未来设计
服务设计
广告策划
公关策划
活动设计
产品设计
网站设计
环境设计
时尚设计
生活方式设计
通信设计
概念设计

3.1.2　2021 DESIGN DAY MARATHON COMPOSITION

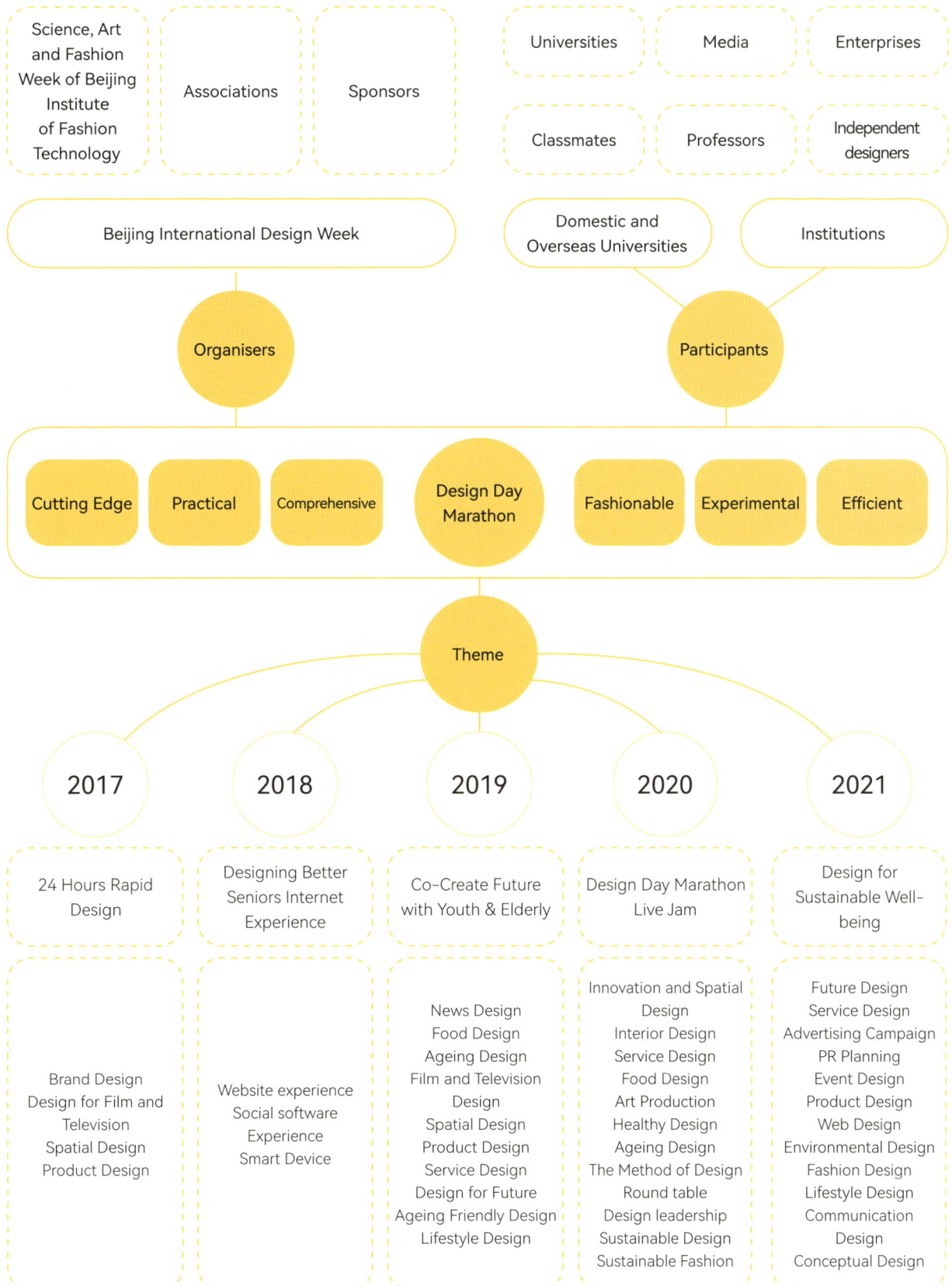

Science, Art and Fashion Week of Beijing Institute of Fashion Technology

Associations

Sponsors

Universities

Media

Enterprises

Classmates

Professors

Independent designers

Beijing International Design Week

Domestic and Overseas Universities

Institutions

Organisers

Participants

Cutting Edge

Practical

Comprehensive

Design Day Marathon

Fashionable

Experimental

Efficient

Theme

2017

2018

2019

2020

2021

24 Hours Rapid Design

Designing Better Seniors Internet Experience

Co-Create Future with Youth & Elderly

Design Day Marathon Live Jam

Design for Sustainable Well-being

Brand Design
Design for Film and Television
Spatial Design
Product Design

Website experience
Social software Experience
Smart Device

News Design
Food Design
Ageing Design
Film and Television Design
Spatial Design
Product Design
Service Design
Design for Future
Ageing Friendly Design
Lifestyle Design

Innovation and Spatial Design
Interior Design
Service Design
Food Design
Art Production
Healthy Design
Ageing Design
The Method of Design
Round table
Design leadership
Sustainable Design
Sustainable Fashion

Future Design
Service Design
Advertising Campaign
PR Planning
Event Design
Product Design
Web Design
Environmental Design
Fashion Design
Lifestyle Design
Communication Design
Conceptual Design

3.1.3　2021 设计马拉松日程

3.1.3　2021 DESIGN DAY MARATHON SCHEDULE

第一步　Step 1

| 策划期 Preparation | 2—4 月 Feb.–April | 活动策划 合作洽谈 | Event planning Negotiation |

第二步　Step 2

| 宣讲期 Announcement | 4—6 月 April–June | 国际院校在线宣讲 国内院校在线宣讲 邀请导师 | Online announcement for international schools Online announcement for domestic schools Inviting tutors |

第三步　Step 3

| 报名期 Pre-Registration | 6—8 月 June–August | 0731 工作坊答疑会 0815 学员报名筛选 0831 确认学员名单 | 0731 Workshop Q&A 0815 Student Pre-registration 0831 Announcing registration result |

第四步　Step 4

| 即兴直播间 Live Jam | 9 月 September | 0911 线上圆桌会议 0912 线上课题宣讲 | 0911 Online round table 0912 Introduction of workshop tasks |

第五步　Step 5

| 工作坊 Workshop | 9 月 September | 0913 学员进入课题组 0919 线上工作坊开始 0923 线上中期检查 1002 线上最终答辩 | 0913 Students' allocation to workshop tasks 0919 Online workshop starts 0923 Intermediate review 1002 Final review |

第六步　Step 6

| 成果展 Workshop Exhibition | 10 月 October | TBA 幸福指南展 | TBA Well-being guide exhibition |

3.1.4 即兴直播间

即兴直播间是个线上主题沙龙，分为线上圆桌会议与线上课题宣讲两部分，这是学员在进入课题组之前认识课题与熟悉导师的最佳渠道。线上专家圆桌会议邀请行业与学术专家对本次主题"可持续福祉"的一系列问题进行讨论与交流，目的是为学员们提供一个理解设计观点与解读案例的学习机会。线上导师邀请全球多个国家的导师讲述他们视角下的设计福祉，并针对其课题的设计方案和后期实践目标进行详细描述和说明。

3.1.4 LIVE JAM

The live jam is an online theme salon, including two parts: an Online Theme Forum and an Online Tutors Roundtable. It is the best platform for the students to get to know the tasks and the tutors before joining any task group. The Online Experts Roundtable will invite professionals and academic experts to discuss a series of issues on the theme of "Sustainable Well-being", to provide students with a learning opportunity to understand design viewpoints and case studies. The Online Tutors Roundtable invites global tutors to talk about their design well-being from their perspectives and to describe and explain their design plans and goals of their tasks in detail.

9 月 11 日 September 11th

时间	内容
10：00—11：00	设计教育学生说：清华+同济+米理（杨叶秋，米兰理工大学） Design Education Students Say: Tsinghua+Tongji+Polimi (Yeqiu Yang, Polytechnic University of Milan, Italy)
11：00—12：00	交互之声——科技与声音的连接（亓梦婕，中国音乐学院） Interactive Voice—The Connection Between Technology and Sound (Mengjie Qi, China Conservatory of Music)
13：00—14：00	如果时尚会说话（郝杰，北京服装学院） If Fashion Can Speak (Jie Hao, Beijing Institute of Fashion Technology)
14：00—15：00	全球背景下的文化项目（Hanny Wijaya，印度尼西亚建国大学） Cultural Projects in Global Contexts (Hanny Wijaya, Binus University, Indonesia)
15：00—16：00	可持续和正念绿色技术实现可持续福祉（Cheeonn Wong，马来西亚多媒体大学） Sustainable and Mindfulness Green Technology Towards Sustainable Well-being (Cheeonn Wong, Multimedia University, Malaysia)
16：00—17：00	自然原则——"瓜田李下"：房前屋后的精神觅处（余春娜，天津美术学院） The Law of Nature—Feel the Fresh Air of Nature (Chunna Yu, Tianjin Academy of Fine Arts)

9 月 12 日 September 12th

时间	内容
10：00—11：00	文化持续力——数字媒介下的非遗研究（赵璐，鲁迅美术学院） Cultural Sustainability—Intangible Heritage Research in Digital Media (Lu Zhao, Luxun Academy Of Fine Arts)
11：00—12：00	养老·社区·服务——基于协作的社区健康服务创新设计（胡鸿，北京工业大学） Pension·Community·Service—Innovative Design Of Community Health Service Based on Collaboration (Hong Hu, Beijing University of Technology)
13：00—14：00	空间激活健康（Anne Farren，澳大利亚科廷大学） Space Activation for Well-being (Anne Farren, Curtin University, Australia)
14：00—15：00	可持续未来，教育创造者（Tanate Panratt，泰国宋卡王子大学） Sustainable Future, Educational Creator (Tanate Panratt, Prince of Songkla University, Thailand)
15：00—16：00	设计关于丽江的诗和远方（王健丰，丽江文化旅游学院） Design Poems and Distances About Lijiang (Jiangfeng Wang, Lijiang Culture and Tourism College)
16：00—17：00	超时空共振：用设计链接传统与时尚（蔡端懿，四川大学） Hyperspace Resonance: Linking Tradition and Fashion with Design (Duanyi Cai, Sichuan University)

3.2　2021 设计马拉松工作坊成果

3.2　2021 DESIGN DAY MARATHON WORKSHOP RESULTS

↘ **3.2.1   课题：15 分钟生活圈：人居环境设计**

3.2.1   TASK: 15 MINUTE CITY: LIVING ENVIRONMENT DESIGN IN THE POST-PANDEMIC ERA

课题说明：

在全球卫生安全、世界经济和国际合作的变化下，我们每个人的生活方式和居住环境发生了改变，人们对"居住社区"也有了新的概念。除了对室内居住空间需求的变化外，人们对社区生活及周边配套设施要求也值得关注。"15分钟生活圈"成为的热门概念，寓意一种崭新"城市乌托邦"的意愿。"15分钟生活圈"是一个都市规划概念，其目标是让城市社区的居民都可以在步行或脚踏车路程可及的范围内，满足衣食住行娱乐等大部分日常所需的服务。此概念首先是由法裔哥伦比亚学者 Carlos Moreno 在 2016 年所提出，之后又由法国巴黎市长安妮·伊达尔戈所推广。此概念可被描述为"回归本地生活方式"。本次工作坊的愿景是重新定义健康、安全的美好人居环境空间，设计的对象是基于每个人身边的 15 分钟环境空间。我们将基于"设计四秩序"（符号—物体—行动—想法）的模型来思考人居环境的系统性解决方案。区别于传统教学的设计"结果导向型"的教学方式，我们更强调设计"思维系统性"的训练。在整个设计过程中以"双钻模型"（发现—定义—发展—交付）为导向，强化训练同学们的以人为本的、系统的、协同的和可视化的"设计思维"。

指导老师：

杨叶秋，意大利米兰理工大学设计学在读博士研究生，英国密德萨斯大学访问学者，同济大学设计创意学院访问学者

Task Description:

Global health security, the world economy and international cooperation have been changed, so have the lifestyle and living environment of each of us. The new concept of a "living community" emerged. In addition to the changes in people's demand for indoor living space, the requirements for community life and infrastructure are also worthy of attention. The "15 Minute City" concept is endowed with a new desire for "Urban Utopia". This project is an urban planning concept aiming at meeting the residents' daily needs, such as food, clothing, housing, transportation, and entertainment within walking or bicycle distance. This concept was first proposed by Carlos Moreno, a French-Colombian scholar, in 2016 and then promoted by the Mayor of Paris, Anne Hidalgo. This concept can be described as "returning to the local lifestyle". This workshop aims to stimulate contemplation regarding the definition of health, safe, and sound livingspace after the pandemic, and all the designs are based on the "15-minute" space of each person. Based on themodel of "Four Orders of Design" (Symbols, Objects, Actions, Thoughts), we will come up with systematic solutions to human settlement. We emphasize the training of designing a "thinking system" instead of the traditional"result-oriented" one."In the whole design process, the "double-diamond model" (Discovery, Define, Development, Delivery) is used as a guide to training students' human-oriented, systematic, collaborative, and visual design thinking.

Tutor:

Yeqiu Yang, a Ph.D. candidate in Design, Polytechnic University of Milan, Italy; visiting scholar at Middlesex University, UK; visiting scholar at the School of Design and Creativity, Tongji University

3.2.1.1 15 分钟生活圈之垃圾系统管理与改造——以方家胡同为例

3.2.1.1 15-Minute Living Circle Management and Renovation of Waste System—Take Fang Jia Hutong As An Example

作者：陈思如、李作鑫、李天翼

By Siru Chen, Zuoxin Li, Tianyi Li

在探索 15 分钟生活圈的理念下，方家胡同的垃圾系统管理与改造项目采取了一种创新的方法，即通过对胡同内拐点的精心设计改造，建立起一个覆盖垃圾投放、回收及再利用的可持续管理系统。这一系统旨在优化居民的日常垃圾处理流程，让垃圾分类和回收变得更加便捷和高效。通过在胡同的策略性位置设置多功能的垃圾分类投放站，不仅提升了垃圾收集的效率，还鼓励了居民参与到垃圾减量和资源回收的环保活动中来。同时，该项目还通过教育宣传提高了居民对于垃圾分类重要性的认识，共同构建一个更加干净、绿色、和谐的生活环境。

In exploring the concept of the 15-minute living circle, the garbage system management and renovation project in Fangjia Hutong adopted an innovative approach by redesigning and transforming the turning points within the hutong to establish a sustainable management system covering garbage disposal, collection, and recycling. This system aims to optimize residents' daily waste handling processes, making waste sorting and recycling more convenient and efficient. By strategically placing multifunctional waste sorting stations at key locations within the hutong, the efficiency of waste collection has been improved, and residents are also encouraged to participate in environmental protection activities focused on waste reduction and resource recovery. Moreover, the project raises residents' awareness of the importance of waste sorting through educational campaigns, collectively creating a cleaner, greener, and more harmonious living environment.

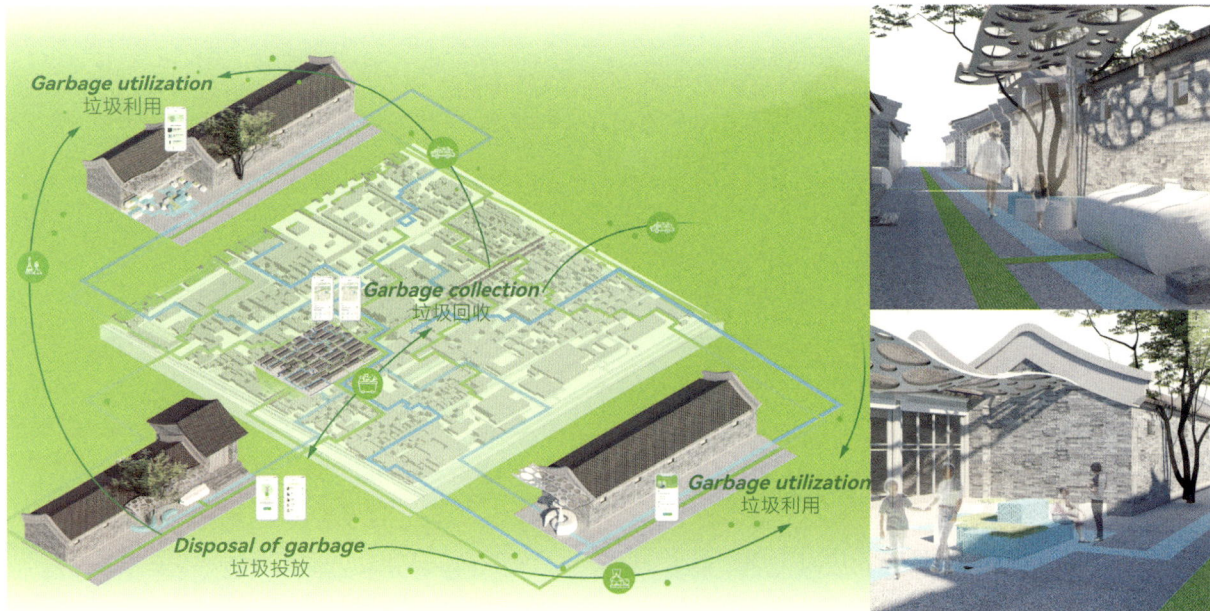

3.2.1.2　社区老年健身产品 + 服务设计

作者：李霜宁、赵晓玉、王世萍、巫佩臻

通过对黑窑厂社区（北京市西城区）的健身和社交需求的设想，使最终成果达到体育与医学的结合、老年人的疾病预防、健身辅助、身体机能训练和社交场所五个功能。您只需要在社区的终端处选择某个或多个节点进行锻炼并成功打卡，即可兑换积分领取奖励。黑窑厂社区老年健身产品 + 服务设计在社区中可以被全面覆盖，提高老年人参与度的同时，也自然形成了社交圈。我们的服务体系依赖于社区的共同维护，而商业模式则选择与相关部门合作，打造黑窑厂网红打卡地。

3.2.1.2　Community Elderly Fitness Product+Service Design

By Shuangning Li, Xiaoyu Zhao, Shiping Wang, Peizhen Wu

Through the assumption of the fitness and social needs of the Heiyaochang community(Xicheng District of Beijing), the results are satisfied: five functions of integration of sports and medicine, disease prevention for the elderly, fitness assistance, physical function training, and social places. You only need to select one or more nodes to exercise at the terminal of the community and check in successfully, then you can exchange points to receive rewards. The design of fitness products and services design for the elderly in the Heiyaochang community can be fully covered in the community, which increases the participation of the elderly and naturally forms a social circle. Our service system relies on the joint maintenance of the community, and the business model chooses to cooperate with relevant departments to create a black kiln factory Internet celebrity check-in place.

3.2.1.3　15 分钟生活圈之儿童公共空间设计：以北京芍药居北里社区为例

3.2.1.3　15-Minute Life Circle for Children's Public Space Design: Taking Beijing Shaoyaoju Beili Community as An Example

作者：陈馨月、李可欣、余旭雯、陈逾

By Xinyue Chen, Kexin Li, Xuwen Yu, Yu Chen

我们为儿童设计了特有的公共空间，在 15 分钟生活圈的基础上，设置了漂流阅读站、滚动乐园、移动绿洲、冲刺巅峰四个节点，以解决在老旧空间内儿童没有娱乐空间的问题。

We have designed a unique public space for children. Based on the 15-minute life circle, we have set up four nodes: Drift Reading Station, Rolling Paradise, Mobile Oasis, and Sprint to the Peak. It is used to solve the problem of children having no entertainment space in the old space.

3.2.1.4 可移动社区购物体系

3.2.1.4 Mobile Community Shopping System

作者：李博然、张嘉琦、方慧娴

By Boran Li, Jiaqi Zhang, Huixian Fang

通过"15分钟生活圈"课题的发散调研，我们充分洞察了北京服装学院周边社区的环境和用户，将老年人需求以及街边小商贩、快递三轮车作为设计灵感来源，提出"可移动社区购物体系"，通过重新设计三轮车并将多个模块连接，形成社区移动商店，以固定时间和区域运营，方便社区生活中老人购物，也帮助商家营利。

According to the diverging research of the "15-minute life circle" task, we have fully gained insight into the environment and users of the surrounding communities of the Beijing Institute of Fashion Technology. The needs of the elderly, street vendors, and express tricycles are the sources of design inspiration, and a "mobile community shopping system" is proposed", by redesigning the tricycle and connecting multiple modules to form a community mobile store, which operates at a fixed time and area, making it convenient for older adults in community life to shop and helping vendors to get rich.

↘ **3.2.2 课题：北京冬奥会游戏化设计**

3.2.2 TASK: GAME DESIGN OF BEIJING WINTER OLYMPIC GAMES

课题说明：

2022北京冬奥会是举世瞩目的体育赛事，举办时正值中国春节假期。为了传播和普及冰雪项目运动的知识，共赏冬奥的魅力，本课题利用微信平台的社交性和传播力，为全民健身、全民共赏的奥运赛事提供服务，打造系列互动公益产品。

指导老师：

宁兵，北京服装学院艺术设计学院数字媒体艺术系副教授

马官正，南京信息工程大学数字媒体艺术系讲师

Task Description:

The 2022 Beijing Winter Olympic Games is a world-renowned sports event during Chinese Spring Festival holiday. To spread and popularize the knowledge of ice and snow sports and enjoy the charm of the Winter Olympics, this task uses the sociality and communication power of the WeChat platform to serve the national fitness and the Olympic events shared by the whole people and create a series of interactive public welfare products.

Tutor:

Bing Ning, Associate Professor of Digital Media Art, School of Art and Design, Beijing Institute of Fashion Technology

Guanzheng Ma, Lecturer of the Department of Digital Media Art, Nanjing University of Information Science and Technology

数字盲盒

作者：李赢、李子一、熊佳玉、张睿、王婷雯、黄雨欣、张萌、王辰雨、黄冬青

　　2022北京冬奥会与中国春节的相遇是中国乃至世界的大事，这个集合了体育精神和团聚欢乐的春节一定会成为中国和世界的永久记忆。北京冬奥数字盲盒 DISCO，是一个游戏化内容服务产品。它以"数字、时尚、有趣"为特色，盲盒游戏机制结合奥运叙事作为框架，是一个以奥运动作舞蹈、赛事实时热点、春节特色文化为内容的创新概念设计。它具有很好的落地性，传播性和可持续性。北京冬奥数字盲盒 DISCO 是由九位青年设计师和两位导师共同创作，我们期待冬奥会开幕时，我们一起开启冬奥盲盒，更期待你加入我们，一起跳出赛事 DISCO！

Digital Blind Box

By Ying Li, Ziyi Li, Jiayu Xiong, Rui Zhang, Tingwen Wang, Yuxin Huang, Meng Zhang, Chenyu Wang, Dongqing Huang

　　The encounter between the 2022 Beijing Winter Olympics and the Chinese Spring Festival is a major event for China and the world. This Spring Festival, which combines sportsmanship and reunion joy, will surely become a long-term memory for China and the world. Beijing Winter Olympics Digital Blind Box Disco is a gamified content service product. It features "digital, fashion, and fun", the blind box game mechanism combined with the Olympic narrative as the framework, the innovative concept design of the Olympic action dance, the real-time hotspot of the event, and the characteristic culture of the Spring Festival. It has good landing, dissemination, and sustainability. Beijing Winter Olympics Digital Blind Box Disco comprises 9 young designers and two mentors. We look forward to opening the Winter Olympics Blind Box together when the Winter Olympics opens, and we look forward to you joining us and jumping out of the event disco together!

Common model
普通款99%

Basic sports edition

冬奥会体育项目版
抽取概率较大
Ordinary model: Winter Olympic Games sports version, with a higher probability of selection

Hide the money
隐藏款1%

Lion dance version

中国传统文化版
抽取概率为 1/100
Hidden section: Chinese traditional culture version, with a probability of 1/100

A rare type
稀有款0.1%

Sports Star Edition

冬奥会体育明星版
抽取概率为 1/1000
Rare model: Winter Olympics sports star version, with a probability of 1/1000

IP +14个运动赛事动作
IP + basic action for 14 sports events

盲盒编号007
Blind Box No. 007

↘ **3.2.3 课题：乐活动态图像设计课程**

3.2.3 TASK: MOTION GRAPHIC DESIGN FOR LOHAS

课题说明：

乐活族是一个重视健康、关爱环保，在消费时以健康、环保、时尚、有机、天然、绿色为主题的群体。在欧美地区，四个人中有一个人是乐活族，仅在美国就有40%的成人，约八千万人是乐活族。中国有三千万乐活族、1亿准乐活族，乐活族强调"健康、可持续的生活方式"。"乐活"是一种环保理念，一种文化内涵，一种时代产物。作为图像创作者，我们将通过可视化的呈现，将这一种贴近生活本源、自然、健康、精致的生活态度传递给大众。

指导老师：

索璐，天津美术学院实验艺术学院动画艺术系讲师

张玥，天津美术学院实验艺术学院动画艺术系讲师

Task Description:

LOHAS is a group of people who pay attention to health and environmental protection and take health, environmental protection, fashion, organic food, nature, and greenness as the theme of consumption. One in four people in Europe and the United States are LOHAS. In the United States, 40 percent of adults, about 80 million people, are LOHAS. There are 30 million LOHAS in China and 100 million people are going to be." LOHAS is an environmental protection concept, a cultural connotation, and a product of the times. As image creators, we will pass this natural, healthy and delicate life attitude to the public through visual presentation.

Tutor:

Lu Suo, Lecturer of the Department of Animation Art, School of Experimental Art, Tianjin Academy of Fine Arts

Yue Zhang, Lecturer of the Department of Animation Art, School of Experimental Art, Tianjin Academy of Fine Arts

3.2.3.1　乐活动态图形设计

3.2.3.1　LOHAS Activity Graphic Design

作者：高昊天、郑子鸣、李萌宇、谷祎楠、冯心雨、柯柔瑄、张泓桦

By Haotian Gao, Ziming Zheng, Mengyu Li, Yinan Gu, Xinyu Feng, Rouxuan Ke, Honghua Zhang

　　LOHAS 是 Lifestyles of Health and Sustainability "健康与可持续生活方式"的缩写，代表一个关注健康与健身、环境保护、个人成长、可持续生活和社会正义的市场细分群体。认同 LOHAS 运动的人们优先考虑那些对自己的健康有益，且对地球环境友好的选择、行为和生活方式。他们倾向于选择那些可以证明拥有积极社会和环境影响的产品和服务，从而推动一个更加可持续和公正的全球社会的形成。乐活族的生活方式聚焦在绿色可持续上面。本作品以此为出发点，创作了种子宝宝、时尚棉棉、杨枝甘露等一系列以绿色健康为主题的人物角色及动画，呼吁人们多多关注绿色可持续生活。

　　LOHAS stands for "Lifestyles of Health and Sustainability", representing a market segment that focuses on health and fitness, environmental protection, personal development, sustainable living, and social justice. Individuals who identify with the LOHAS movement prioritize choices, behaviors, and lifestyles that are beneficial to their own health and also friendly to the environment. The LOHAS lifestyle focuses on green sustainability. Taking this as a starting point, this work has created a series of characters and animations with the theme of green health, such as seed babies, fashionable cotton, sheep branch nectar, etc., calling on people to pay more attention to green and sustainable life.

3.2.3.2 乐活星计划

3.2.3.2 LOHAS Planet Program

作者：花杰、林杰、高洁、王佳慧、连婕、庄皓勋、曹畅

By Jie Hua, Jie Lin, Jie Gao, Jiahui Wang, Jie Lian, Haoxun Zhuang, Chang Cao

20XX年，随着人类科技的进步和社会的发展，地球的环境危机越来越严重，人们的生活质量也越来越低；七位人类先驱成立"乐活有限公司"并隆重推出乐活"星"计划，阿伍被选中参与该计划，即将前往乐活星球，他会遇到什么呢？

In 20XX, with the development of human technology and society, the environmental crisis of the earth is becoming more and more serious, and people's quality of life is also getting lower and lower; seven human pioneers established "LOHAS Co., Ltd. "and solemnly launched". Awu has been selected to participate in the "LOHAS Planet Program" and will soon be heading to the LOHAS planet. What will he encounter there?

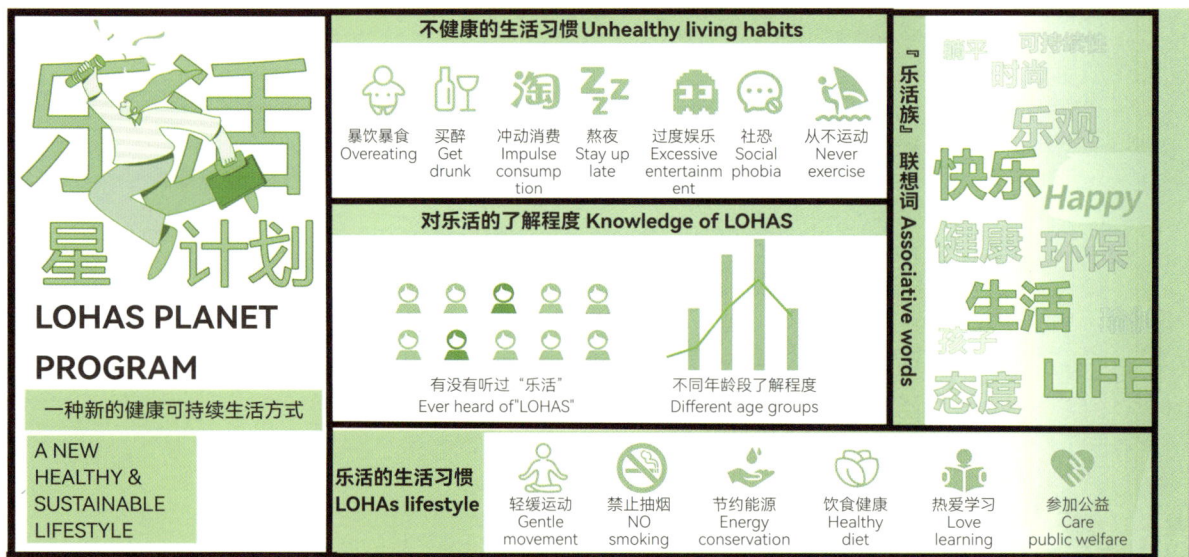

3.2.3.3 乐活动态图像设计

3.2.3.3 LOHAS Motion Image Design

作者：何茜、张杨、万春彤

By Xi He, Yang Zhang, Chuntong Wan

乐活社区里住着乐活一族，海报内容展现了乐活族的方方面面。海报内容以衣、食、住、行为主，分别体现了乐活族在衣、食、住、行方面的生活方式和生活理念，倡导可持续发展的健康生活。

The LOHAS family lives in the LOHAS community, and the content of the poster shows all aspects of the LOHAS family. The content of the poster is mainly based on clothing, food, housing, and behavior, which respectively reflects the lifestyle and life philosophy of the LOHAS people in terms of clothing, food, housing, and transportation, and advocates a sustainable and healthy life.

LOHAS motion image design
乐活动态图像设计

The draft design
草案设计

↘ **3.2.4 课题：乐龄直播**

3.2.4 TASK: HAPPY AGE LIVE BROADCAST

课题说明：

根据世界卫生组织预测，全球银发经济规模将于 2025 年达到新台币 1.122 兆元，是 21 世纪最具潜力的产业之一。个人化直播之所以吸引人，在于能让观众产生一种进入了主播生活之中的感受。直播让人"进入在场"的技术特性，让无聊的、甚至觉得孤单的人们，体验到进入他人生活的"真实感受"。高手在民间，这群乐龄长者想分享的东西很多，但对于复杂的直播技术掌握有难度，省去繁复的后序制作程序和各种拍摄装备之后，是否能为乐龄族打开全民直播的窗口？

指导老师：

黄文宗，中原大学商业设计系副教授，设计学院博士生导师

Task Description:

According to the World Health Organization's forecast, the global silver-hair market is one of the most promising industries, reaching NT$1,122trillion in 2025. What makes personalized live broadcast intriguing is that it gives viewers the feeling of entering the anchor's personal life. The technical characteristics of live broadcasting allow people to "enter the presence", introducing people who are bored or even lonely to the true feeling of life by entering others. There are many things that this group of senior citizens want to share, but it is difficult to master the complicated live broadcast technology. By eliminating the complicated post-production processes and various shooting equipment, is it possible to open the window of the national live broadcast for the seniors?

Tutor:

Wenzong Huang, Associate Professor, Department of Business Design, Chung Yuan Christian University, Doctoral Supervisor of the School of Design

乐龄直播

Happy Age Live Broadcast

作者：梁静雯、梅笑雪、乜可昕、彭汉婷、潘苗、邬媛媛、吴宇凡

By Jingwen Liang, Xiaoxue Mei, Kexin Nie, Hanting Peng, Miao Pan, Yuanyuan Wu, Yufan Wu

收音机是老一辈人共同的记忆，但现有的收音机未能适应当代数字化发展，所以我们以收音机为原型，以老年人的兴趣爱好为驱动，针对智能机操作复杂、不易上手等问题，做出了以下改进：保留电台功能，并建立语音直播聊天室，使有共同爱好的老人可以相互交流，同时利用大数据智能匹配老人兴趣，推送相关聊天室。我们希望通过这种方式解决老年人想要去交流和分享但苦于没有渠道的问题，以更加适龄化的产品去帮助老年人更好地融入当今数字化社会。这符合了联合国2030年可持续发展的第三项："确保健康的生活方式，促进各年龄段所有人的福祉"要求。未来我们的产品将会走向国际化，继续增加全世界的乐龄一族的福祉。

The radio is the common memory of the older generation, but the existing radios have not been able to adapt to contemporary digital development, so we take the radio as the prototype, driven by the hobbies of the elderly, and make the following problems for the complex operation of smartphones and difficult to use. Improvement: Retain the radio function and establish a voice live chat room so that elderly people with common hobbies can communicate with each other, and at the same time use big data to intelligently match the interests of that the elderly and push relevant chat rooms. In this way, we hope to solve the problem that the elderly want to communicate and share but suffer from a lack of channels and use more age-appropriate products to help the elderly better integrate into today's digital society. This is in line with the UN's 2030 Sustainable Development Item 3: "Ensuring healthy lives and promoting well-being for all at all ages". In the future, our products will go international and continue to increase the well-being of the elderly around the world.

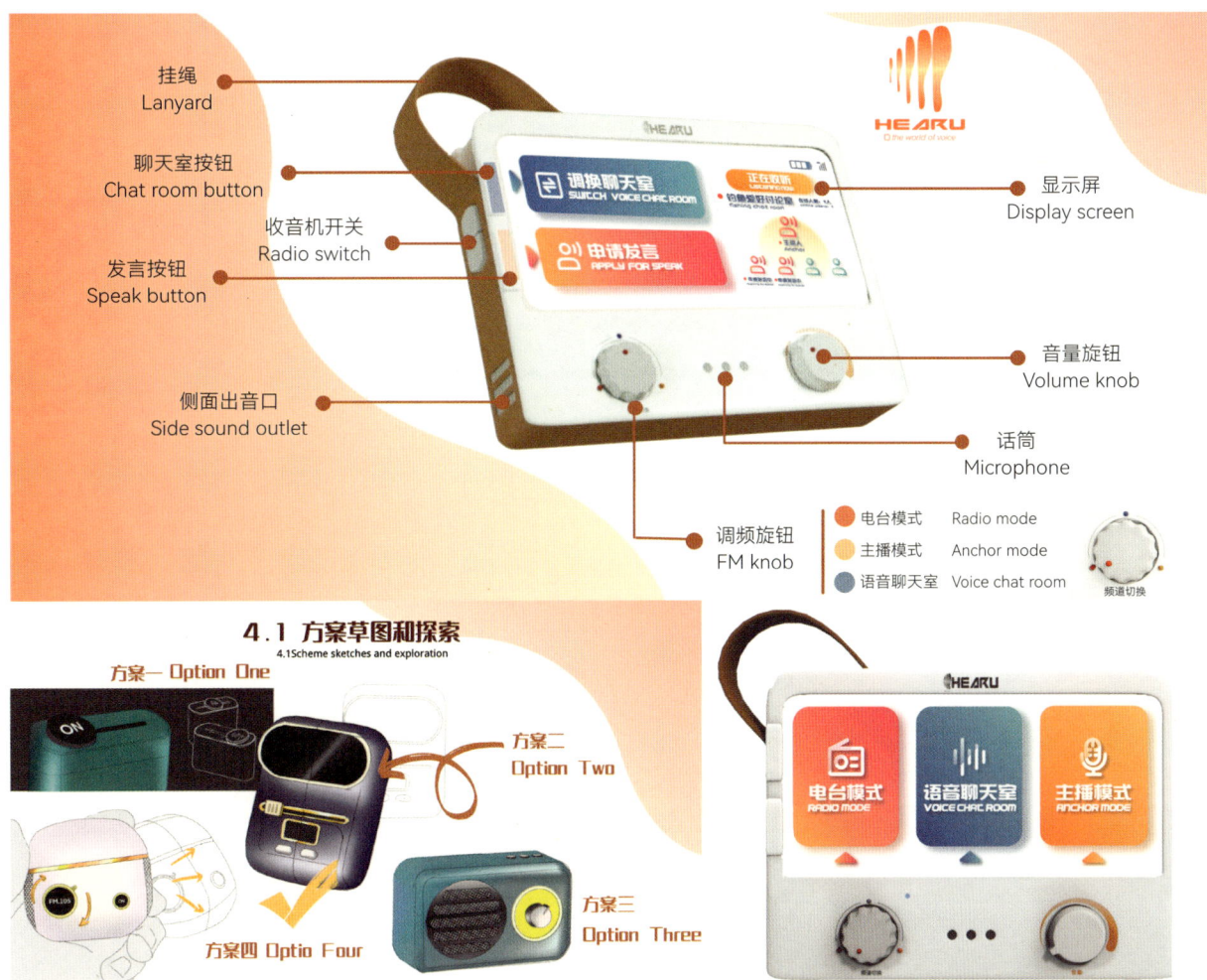

挂绳
Lanyard

聊天室按钮
Chat room button

收音机开关
Radio switch

发言按钮
Speak button

侧面出音口
Side sound outlet

显示屏
Display screen

音量旋钮
Volume knob

话筒
Microphone

调频旋钮
FM knob

电台模式 Radio mode
主播模式 Anchor mode
语音聊天室 Voice chat room

频道切换

4.1 方案草图和探索
4.1Scheme sketches and exploration

方案一 Option One

方案二 Option Two

方案三 Option Three

方案四 Optio Four

↘ **3.2.5　课题：羌绣的图案设计创新与时尚应用**

3.2.5　TASK: PATTERN DESIGN INNOVATION AND FASHION APPLICATION OF QIANG EMBROIDERY

课题说明：

羌绣图案寓意深刻，色彩明快，兼具美观与实用功能，反映出羌族人民的造物智慧与美好生活憧憬。作为设计师，我们将通过发掘羌族文化，研究羌绣的技艺特征与视觉语言，结合当下审美文化与生活需求进行图案设计创新，并以时尚产品与数码媒介为载体探索其转化与应用途径，以设计为媒连接民族与世界、传统与时尚，为多元审美文化贡献力量；同时，通过视觉设计方法带动羌文化中"万物有灵"的自然观念融入当下生活方式，响应"人与自然生命共同体"构建。

Task Description:

The patterns of Qiang embroidery have precondemning and bright colors. They have both beautiful and practical functions, reflecting the creative wisdom and vision of a better life for the Qiang people. As designers, we will explore the Qiang culture, study the technical characteristics and visual language of Qiang embroidery, innovate the pattern design in combination with the current aesthetic culture and life needs, explore its transformation and application ways with fashion products and digital media as the carrier, and use design as the media to connect the nation and the world, tradition and fashion, to contribute to the diversified aesthetic culture; At the same time, through the visual design method, the natural concept of "everything has spirit" in Qiang culture is integrated into the current lifestyle, in response to the construction of "life community between man and nature".

指导老师：

蔡端懿，四川大学艺术学院讲师

许亮，副教授，四川大学艺术学院设计与媒体艺术系主任

Tutor:

Duanyi Cai, Lecturer at College of Arts, Sichuan University

Liang Xu, Associate Professor, Dean of the Department of Design and Media Art, College of Arts, Sichuan University

3.2.5.1　羌族文字的设计创新和推广应用

作者：赵跃、陈葳、杜昌林、张逸帆、孙坦、牟芳芳、麦扬、许晓怡

　　羌族具有悠久的历史，但是由于古代的羌族没有文字，这个民族几千年的文化也变得扑朔迷离了起来。新中国成立以来，在政府的努力下，羌族有了自己的文字，羌文能更好地记录羌族的文化，具有重要的作用。但是，由于种种原因，羌族文字没有得到大规模的推广和使用。我们将羌族文字和羌绣图案、色彩进行了创新设计，借助羌绣的图形语言、丰富内涵及视觉感染力来表达羌族文字的含义与意境；同时，将羌族文字与汉字相结合，通过异质同构的图形创意方式进行文字创新设计，借助汉字广泛的应用度和强大传播力，辅助羌文字的认知与推广。

3.2.5.1　Design Innovation and Promotion of Qiang Characters

By Yue Zhao, Wei Chen, Changlin Du, Yifan Zhang, Tan Sun, Fangfang Mou, Yang Mai, Xiaoyi Xu

　　The Qiang people have a long history, but since the ancient Qiang people did not have written characters, the culture of this nation for thousands of years has become confusing. Since the founding of the People's Republic of China, with the efforts of the government, the Qiang people have their own writing, and the Qiang language can better record the culture of the Qiang people and has an important position. However, due to various reasons, the Qiang script has not been widely promoted and used. We have innovatively designed Qiang characters and embroidery patterns and colors and expressed the meaning and artistic conception of Qiang characters with the help of the graphic language, rich connotation, and visual appeal of Qiang embroidery. The innovative design of characters is carried out in an isomorphic graphic creative way. With the help of the wide application and strong communication power of Chinese characters, it assists the recognition and promotion of Qiang characters.

3.2.5.2 万物有灵——羌绣创作体验空间

3.2.5.2 Animism—Qiang Embroidery Creation Experience Space

作者：向根玉、张俊彦、燕阳、陈浩立、朱思嘉、魏育星、郑高厅

By Genyu Xiang, Junyan Zhang, Yang Yan, Haoli Chen, Sijia Zhu, Yuxing Wei, Gaoting Zheng

我们以数码媒介为载体，运用交互设计的方法与逻辑和现代时尚视觉语言，设计了一组动态交互装置。如通过人拍手动作，屏幕上的白石互击迸发出羌绣中的太阳花纹样，来表现白石信仰的起源；通过人拍羊皮鼓动作，借用了羌绣针法设计而成的小人随之跳锅庄，表现羌族的自然崇拜；羌绣代表纹样野牡丹跟随路径绽放与聚合，表现"万物有灵"观念；我们设计了一个线上互动，参与装置互动的朋友可以扫码并根据自己的愿望获得一个独有的组合羌绣吉祥图案，也可以把这个祝福分享给没办法来现场的亲朋好友。

Taking digital media as the carrier, we designed a set of dynamic interactive installations using the method and logic of the interactive design and modern fashion visual language. For example, through the action of clapping hands, the white stones on the screen hit each other and burst out the sun pattern in Qiang embroidery, to express the origin of the belief in Baishi; The jumping Guo Zhuang represents the nature worship of the Qiang people; Qiang embroidery represents the pattern of wild peony blooming and gathering along the path, expressing the concept of "animism"; we have designed an online interaction, and friends who participate in the interaction of the installation can scan the code and follow their preferences. Wish to get a unique combination of Qiang embroidery and auspicious patterns, and you can also share this blessing with relatives and friends who can't come to the scene.

3.2.5.3　羌绣的图案设计创新与时尚应用

作者：张笑蕾、夏丽茹、赵一然

文化商品化带来了时尚产品、时尚产业、时尚经济。而时尚产业作为引领世界产业发展的最重要行业之一，体现着一个国家在文化、科技、创新等方面的软实力。羌族文化本身具有稀有艺术价值和产品价值，值得深挖。专业的传统积淀和独特的艺术审美，是天然的产品"护城河"和"防火墙"，而传统文化技术与现代生活创意性的结合，能帮助创造出实用与艺术兼有的价值含量更高的产品。

3.2.5.3　Pattern Design Innovation and Fashion Application of Qiang Embroidery

By Xiaolei Zhang, Liru Xia, Yiran Zhao

Cultural commodification brings fashion products, the fashion industry, and the fashion economy. As one of the most important industries leading the world's industrial development, the fashion industry reflects a country's soft power in culture, technology, innovation, and other aspects. The rare artistic value and product value of Qiang culture itself can be fully explored. Professional traditional accumulation and unique artistic aesthetics can be said to be natural products "moats" and "firewalls". Combining traditional cultural techniques with modern life creativity can help create products with higher value that are both practical and artistic.

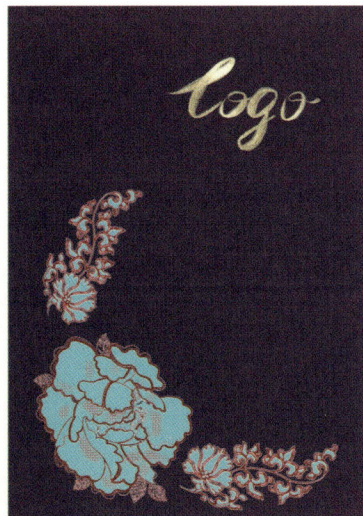

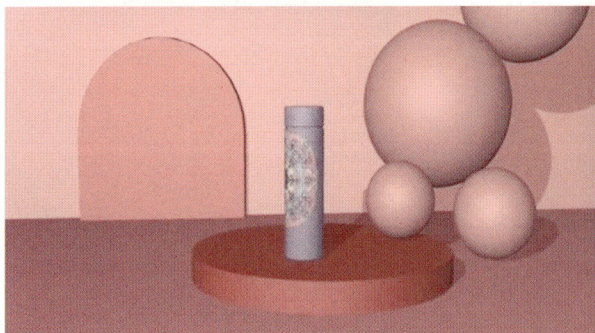

↘ 3.2.6 课题：文化持续力——数字化非遗研究

3.2.6 TASK: CULTURAL SUSTAINABILITY—DIGITAL INTANGIBLE HERITAGE RESEARCH

课题说明：

旨在更好地传承与创新非遗，获得新的表现力和生命力，让越来越多的年轻人关注非遗、感受传统文化之美，了解中国故事，提升青年人对家乡的自豪感和文化自信。课程将通过对三个非遗项目——剪纸、皮影、二十四节气的深入研究，最终产出可以帮助非遗在新媒体语境下高效传播的设计作品，包括且不限于：文创产品、影像、IP形象、表情包和动态H5宣传页面。

Task Description:

In order to better inherit and innovate intangible cultural heritage, gain new expressiveness and vitality, let more and more young people pay attention to intangible cultural heritage, feel the beauty of traditional culture, understand Chinese stories, and enhance the pride and culture of young people in their hometown confidence. The course will through in-depth research on three intangible cultural heritage projects, paper-cutting, shadow puppetry, and twenty-four solar terms, and finally produce design works that can help intangible cultural heritage to efficiently spread in the context of new media, including but not limited to cultural and creative products, Image, IP image, emoji package, dynamic H5 promotion page.

指导老师：

赵璐，教授，博士生导师，鲁迅美术学院副院长

毕卓异，鲁迅美术学院讲师

Tutor:

Lu Zhao, Professor, Doctoral Supervisor, Vice President of LuXun Academy of Fine Arts

Zhuoyi Bi, Lecturer at LuXun Academy of Fine Arts

3.2.6.1　尘封的艺术和视觉扩张——傩戏面具在数字媒体下的叙事表达

3.2.6.1　Dusty Art and Visual Expansion—The Narrative Expression of Nuo Opera Masks in Digital Media

作者：王啸林、王然、居丽洁、刘苡辰、李晨华、江依诺

By Xiaolin Wang, Ran Wang, Lijie Ju, Yichen Liu, Chenhua Li, Yinuo Jiang

《傩·五猖》这个作品是根据傩戏中的"跳五猖"这一出戏为基本，加以研究创新，通过数字媒介让它展现在更多人的面前。五猖是当地农民在每逢元旦、春节和重大节庆活动中一项不可或缺的重头戏。但随着文化环境的消失和多元文化的冲击，傩戏濒临失传。所以我们通过IP形象、动画、表情等形式创作这个作品，让更多的人看到我们的傩文化，让我们的傩戏能继续传承下去。

The work *Nuo, Five Devils* is based on the dance of five rampages in Nuo Opera. Plus, with our research and innovation through digital media, we make it appear in front of more people. Five rampages is an indispensable highlight for local farmers in every New Year, Spring Festival, and major festivals. However, with the disappearance of the cultural environment and the impact of multiculturalism, Nuo Opera is almost lost. Therefore, we made this work through IP images, animations, expressions, etc., so that more people can see our Nuo culture and our Nuo Opera can continue to be passed down.

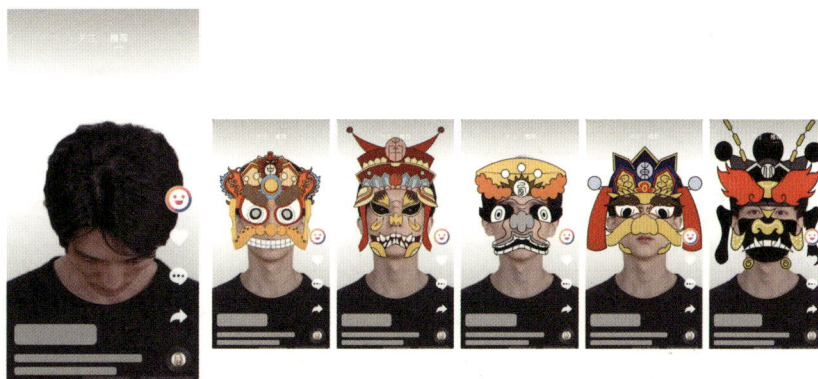

2021 DESIGN DAY　Offline derivatives　线下衍生品

Fun five Devils mask
趣味五猖面膜

2021 DESIGN DAY　Offline derivatives　线下衍生品

Five Devils masks
五猖口罩

3.2.6.2 非遗"活"起来，瓦猫"伴"身边

3.2.6.2 The Intangible Cultural Heritage "Lives," and the Tile Cat Is "Accompanied" by the Side

作者：孙诗涵、彭心语、黄宝双、陈妮

By Shihan Sun, Xinyu Peng, Baoshuang Huang, Ni Chen

本作品深入挖掘瓦猫背后独特的文化内涵，结合瓦猫乖张独特的外形，打造了一款治愈系瓦猫养成类APP，利用当今互联网传播速度快，范围广的优势，将瓦猫非遗推向大众。在作品中，我们加入了全息投影交互技术，让用户与瓦猫互动时更加投入；在线下宣传方面，我们设计策划了主题文创和3D娃娃展览，让瓦猫文化逐渐融合到大众的生活中，从而将这一逐渐消失的非遗推向世界。

This work deeply excavates the unique cultural connotation behind the tile cat and combines the unique appearance of the tile cat to create a healing APP for the development of the tile cat. Taking advantage of the fast speed and wide range of the Internet today, the tile cat is an intangible cultural heritage. Culture to the masses. In the works, we have added holographic projection interactive technology to allow users to be more engaged when interacting with the tile cat. In terms of offline publicity, we have integrated the theme of cultural creation and 3D doll exhibition, so that the tile cat culture can gradually be integrated into the lives of the public. In this way, this gradually disappearing intangible cultural heritage will be introduced to the world.

3.2.6.3　花中道——传统中式插花游戏

3.2.6.3　Flower Middle Road—Traditional Chinese Flower Arrangement Game

作者：王玉洁、王萌、田悦、温玉婷、任碧莹

By Yujie Wang, Meng Wang, Yue Tian, Yuting Wen, Biying Ren

这是一个关于中式插花的线上互动游戏，用户可以按照传统插花的流程根据引导完成一幅插花作品。中式插花崇尚自然简约之美，善用线条造型和不对称构图营造诗情画意的世界，能够充分表现出中华民族特色和传统中国人的审美意识。商业化是最好的保护，使用是最好的传承，分享是最好的传播。后期可以与花店联名，定制个性化花礼，赋予中式插花新生命，跨越时空和距离为想念的人送上一份别出心裁的祝福。

This is an online interactive game about Chinese flower arrangements. Users can follow the traditional flower arrangement process to complete a flower arrangement work according to the guidance. Chinese flower arrangements advocate the beauty of natural simplicity, making good use of line shape and asymmetrical composition to create a poetic world, fully showing the characteristics of the Chinese nation and the aesthetic awareness of traditional Chinese people. Commercialization is the best protection, use is the best inheritance, and sharing is the best dissemination. In the later stage, you can co-brand with a flower shop, customize personalized flower gifts, give new life to Chinese flower arrangements, and send an ingenious blessing to those you miss across time, space, and distance.

什么是中式传统插花？
What is a traditional Chinese flower arrangement?

中国传统插花艺术崇尚自然简约之美，善于用线条造型和不对称构图营造诗情画意的境界，充分表现出中华文化的民族特色和传统中国人的审美意识。

The traditional Chinese art of flower arranging is based on the beauty of natural simplicity and the use of linear shapes and asymmetrical compositions to create a poetic and picturesque realm, fully expressing the national characteristics of Chinese culture and the aesthetic sense of the traditional Chinese.

| 桂花 | 菊花 | 兰花 | 鸢尾草 | 萱草花 | 竹子 | 水仙花 | 玉兰 |
| Osmanthus | Chrysanthemum | Orchids | Iris | Daylily | Bamboo | Daffodils | Yulan |

↘ **3.2.7 课题：文化交互传播场景中的用户体验设计**

3.2.7 TASK: USER EXPERIENCE DESIGN IN CULTURAL INTERACTIVE COMMUNICATION SCENARIOS

课题说明：

　　文博场景、图书场景、文艺鉴赏场景和历史资料展示场景（博物馆）中的文化体验用户交互过程被定义为"文化数字交互"，不同场景下的文化传播与用户体验流程在交互行为目的与特征方面具有显著差异，其主要表现为文化内容的广泛性与多样性以及场景建构方法的丰富性。而影响文化记忆传播评价的另一个因素是用户的文化背景、学习方式和互动偏好的个体特征，这一现象为用户体验研究介入文化传播与文化产品研发提供了创新思路。

指导老师：

　　王可，南京艺术学院工业设计学院讲师

Task Description:

The cultural experience user interaction process in cultural scenes, book scenes, literary appreciation scenes, and historical data display scenes (museums) is defined as "cultural digital interaction". The cultural communication and user experience processes in different scenes are based on the purpose of different scene interaction behaviors. It is significantly different from the characteristics, which are mainly manifested in the breadth and diversity of cultural content and the richness of the method of scene construction. Another factor that affects the evaluation of cultural memory communication is the individual characteristics of the user's cultural background, learning style, and interaction preference. This phenomenon provides an innovative way for user experience research to intervene in cultural communication and cultural product development.

Tutor:

Ke Wang, Lecturer, School of Industrial Design, Nanjing University of the Arts

皮影戏文化主题的儿童互动体验设计

Children's Interactive Experience Design on The Theme of Shadow Play Culture

作者：张祎明、沈妍菲、方艺舒、韩雨和、王清慧、张译丹、梁文楷

By Yiming Zhang, Yanfei Shen, Yishu Fang, Yuhe Han, Qinghui Wang, Yidan Zhang, Wenkai Liang

"皮影戏文化主题的儿童互动体验设计"是基于文旅、博物馆等文化空间场景，以皮影为表现形式，输出儿童美育游戏的产品服务系统。该设计旨在为用户提供互动型文创体验，在挖掘皮影戏戏曲艺术表现形式的同时，将儿童对于非物质文化遗产知识的被动科普转化为主动学习。互动设计以《山海经》为脚本，用故事化的方式引导儿童进行游戏互动，增强儿童在视觉、听觉、触觉等方面的多感官体验，在传播非物质文化遗产的同时，提升游戏价值所带来的体验反馈。

"Children's Interactive Experience Design on The Theme of Shadow Play Culture": It is a product service system that outputs children's aesthetic education games based on cultural space scenes such as cultural tourism and museums, using shadow play as a form of expression. The design is intended to provide users with an interactive cultural and creative experience, transforming children's passive popularization of intangible cultural heritage knowledge into active learning while exploring the artistic expression of shadow puppetry and opera. The interactive design takes *Shan Hai Jing* as the script, guides children to interact with games in a story-based way, enhances children's multi-sensory experience in sight, hearing, touch, etc., and enhances the value of games while disseminating intangible cultural heritage. Experience feedback.

交互场景　Interactive Scene

文创纪念品 Creative Cultural Souvenirs

皮影形象呼吸灯 Shadow image breathing lamp

↘ **3.2.8 课题：色彩实验训练**

3.2.8 TASK: COLOUR EXPERIMENTAL TRAINING

课题说明：

对于一个设计师来说擅长灵活使用色彩，产生预期效果，是一个极大的优势。对不同颜色的组合和应用，不但需要理论知识，而且需要设计师有足够的视觉敏感度，以将传播需求转化为色彩语言。色彩与人们的情感紧密连接，并且会与"设计"（物体、空间、服装、电子产品）产生强烈而直接的碰撞。掌握这一强大的语言，探索色彩对情感与行为的作用力，最终有益于设计师掌握基本的设计技巧。

指导老师：

Lia Vilahur，教授，西班牙赫罗纳大学媒体学院国际合作处主任，艺术家

Task Description:

For a designer is an advantage to domain the color language to have the expected impact on his audience. The skill of successfully combining colors and using them in different design fields needs not just the knowledge of theory but also the capacity to identify communication needs and have the eye and sensibility to translate them in correct color usage. Color communicates emotions and has a strong and direct impact on the response and interaction with the "designs" (objects/spaces/clothes and also digital productions). Mastering this powerful communication skill and discovering how it influences emotions, feelings, and behaviors will, in conclusion, help to develop basic design skills.

Tutor:

Lia Vilahur, Professor, Dean of International Cooperation Office, School of Media, University of Girona, Artist

欢乐的精神

Spirit of Joy

作者：林炜娟、Yunhee Min、As-Saiyidah Nafisah binti Halis、Seoyoung Park、Suhyean Choi

By Weijuan Lin, Yunhee Min, As-Saiyidah Nafisah binti Halis, Seoyoung Park, Suhyean Choi

颜色对人们来说非常重要，使用正确的颜色可以改变人们在不同环境中的心情，物品颜色也可以让使用者有不一样的感受。在设计作品中我们使用了生活中会出现的颜色、色调，将收集到的色调拼凑成美丽的图纹，之后将导师帮我们矫正过的色调套用在我们友好空间的公共设施上，不管什么样的年龄都能在年龄友善椅上有美好的相遇；改变以往食物银行的色彩，让在此空间的人们都能有正确的饮食资讯并取得足够的食物；友善照明设备、喷泉可以让人们与当地自然、环境更有联结。

Color is very important to people. Using the correct color can change people's moods in the environment. The color of items can also give users different feelings. In the design works, we use the colors and tones that will appear in life. Piece together the collected shades into beautiful patterns, and then we apply the shades corrected by our tutor to the public facilities in our friendly space. No matter what age, the age-friendly chair can have a beautiful look on this chair. Encounter and change the color of the previous food bank so that people in this space can have correct dietary information and enough food, friendly lighting equipment and fountains can make people more connected with the local nature and environment.

↘ 3.2.9 课题：数字疗法的福祉

3.2.9 TASK: DIGITAL THERAPEUTICS FOR WELL-BEING

课题说明：

此项任务的主题是数字治疗（Digital Therapeutics，DTx）。数字治疗联盟将DTx定义为"基于高质量软件程序驱动，为患者提供循证医学治疗干预，用来预防、管理或治疗医学紊乱或疾病"的产品。医疗保健研究人员和设计师在体验设计、用户参与、参与度和交互系统评估方面面临许多类似问题。例如，越来越多的人对应用与人机交互相关的方法来提高患者对医疗保健系统的参与度这一议题感兴趣。我们将讨论在医疗保健和人机交互融合过程中出现的问题、困难和机遇，重点聚焦于数字治疗。最后，此项任务将会为学生提供一个机会，让他们对创造革命性的研究成果产生新的认识。

指导老师：

Hyungsin Kim，韩国国民大学技术设计研究院智能体验设计系副教授

Task Description:

The task topic is Digital Therapeutics (DTx). DTx is defined by digital therapeutics' alliance as products that "deliver evidence-based therapeutic interventions to patients that are driven by high-quality software programs to prevent, manage, or treat a medical disorder or disease". Healthcare researchers and designers confront many similar issues in terms of experience design, user participation, engagement, and interactive system evaluation. For example, there is a growing interest in applying HCI-related methodologies to enhance patient engagement in healthcare systems. We will discuss issues, difficulties, and opportunities arising in the convergence of healthcare and HCI with a focus on digital therapeutics. Ultimately, this task will provide students the opportunity to shed new light on creating revolutionary research.

Tutor:

Hyungsin Kim, Associate Professor, Department of Intelligent Experience Design, Institute of Technology Design, Kookmin University, Korea

数字疗法的福祉

The Well-Being of Digital Therapy

作者：侯馨月、朱洪萱、刘祥铭、唐楚哲

By Xinyue Hou, Hongxuan Zhu, Xiangming Liu, Chuzhe Tang

Hyperfood是一款针对年轻高血压患者的个性化治疗服务。我们基于云端数据技术，通过整合数字技术、医疗专家、医药公司、食物供应商和外卖平台，构建了从诊疗到日常生活餐饮管理的服务闭环。以生活方式引导为切入点，为青年高血压患者提供便携、智能化血压记录流程、个性化医疗计划和健康化的饮食生活解决方案，解决单次医院测量血压波动误差大、药物治疗方案调整烦琐、饮食数据难追踪等问题，帮助年轻患者与医生共同管理患者的日常血压，轻松吃出健康生活。

Hyperfood is a personalized treatment service for young hypertensive patients. Based on cloud data technology, we have built a closed loop from diagnosis and treatment to daily catering management to service by integrating digital technology, medical experts, pharmaceutical companies, food suppliers and takeaway platforms. Taking lifestyle guidance as the entry point, it provides young hypertensive patients with a portable, intelligent blood pressure recording process, personalized medical plans, and healthy diet and living solutions, and solves the problem of large fluctuations in blood pressure measurement in a single hospital and cumbersome adjustment of drug treatment plans. It helps young patients and doctors manage their daily blood pressure together and eat a healthy life easily.

CHYPERFOOD
"千人千面"的服药故事
Personalized medication plan

CHYPERFOOD
改善血压，脚踏"食"地
Reshaping a healthy diet

真心为你，献上万千健康佳肴
Hundreds of healthy diet recipes.
eat easily and eat healthily

CHYPERFOOD

病历信息导入
Medical Record Information

↘ **3.2.10 课题：Semar —— 生命、希望和未来的代表**

3.2.10 TASK: INTERPRETING SEMAR AS THE REPRESENTATION OF LIFE, HOPE, AND FUTURE

课题说明：

本课题希望鼓励参与者在这段使我们的生活发生巨大改变的艰难时期（自我隔离、身体疏远、居家办公等），探索和发现生活平衡（身体和心理）的概念和语境。参与者也可以找到 Semar 与自己国家本土神话和民间故事之间的联系，并将其阐述于艺术和设计视角下当代生活的平衡与福祉。

指导老师：

Hanny Wijaya，副教授，印度尼西亚建国大学国际项目合作处处长

Task Description:

The task would like to encourage participants to explore and discover the concept and context of life balance (physically and mentally) during this hard time (self-quarantine, physical distancing, work from home, etc.) that has changed our lioes considerably. They also can try to find the connection of Samar's character contextually with mythology or folk stories from their home countries and interpret it into the contemporary context of life balance and well-being from an art and design perspective.

Tutor:

Hanny Wijaya, Associate Professor, dean of International Project Cooperation Office, BINUS University

3.2.10.1 共同生活，共同创造

3.2.10.1 Co-Living, Co-Creation

作者：Mila Savitri, Friska Amalia, Siti Chadijah

By Mila Savitri, Friska Amalia, Siti Chadijah

共同生活、共同创造系统是一种弹出、移动结构，以创意中心设施的形式为城市中的老年人提供更高的生产力，并与他们的社区相处。Semar 是 Punakawan 故事中的人物，我们从 Semar 身上学到了很多价值观——智慧、保护、培育、乐趣和人性。我们以无限的形式重新诠释 Semar，以转化为生命的闭环循环。共同生活和共同创造代表了这些价值观，因为共同生活空间给我们带来了合作和协作的精神。

Co-living, Co-creation System is a kind of pop-up/mobile structure in the form of a creative hub facility for the elderly in the city to be more productive and to get along with their community. From Semar as a figure in the Punakawan story, we learn so many values, suchas wisdom, protecting, nurturing, fun, and humanity. We are re-interpreting Semar in an infinity form, which can be translated into the closed-loop cycle of life. Co-living and co-creation represent those values since co-living space brings us the spirit of Gotong Royong (cooperation) and collaboration.

3.2.10.2　将 Semar 解读为生命、希望和未来的代表

3.2.10.2　Interpreting Semar as The Representation of Life, Hope, and Future

作者：Siti Aisyah Fd、Kadek Ngukuhin Wijaya、Nisriina Salmannida、Zefa Christiansa Patricia Malau

By Siti Aisyah Fd, Kadek Ngukuhin Wijaya, Nisriina Salmannida, Zefa Christiansa Patricia Malau

快节奏的生活方式是我们目前在现代社会中所经历的现实。不幸的是，生活在快节奏的社会中会引发压力反应，并可能导致生活质量下降。我们将 Semar 角色背后的哲学与接地技术相结合，构建了一个装置，观众可以从中体验到雅加达商业区中心的轻松氛围。在装置内部，参与者可以坐下来，在忙碌的日子里稍作停顿。在休息区内，将有镜子帮助参与者反思并与自己和解。

A fast-paced lifestyle is a reality that we currently experience in the modern world as a society. Unfortunately, living in a fast-paced society can trigger stress responses and could lead to a low quality of life. The philosophy behind Semar's character and the grounding technique were combined to build an installation from which the audience can experience a relaxing ambience in the middle of a business district in Jakarta. Inside the installation, the participant can sit down and take a moment to pause amidst the hectic days. Inside the sitting area, there will be mirrors to help the participants reflect and make peace with themselves.

↘ 3.2.11 课题：新理念下的新设计，传播新概念模型和新思维结构的创造性工具

3.2.11 TASK: NEW DESIGN FOR A NEW PHILOSOPHY, CREATIVE TOOLS TO DISSEMINATE AND TRANSMIT NEW CONCEPTUAL MODELS AND NEW STRUCTURES OF THOUGHT

课题说明：

今天，后数字社会比以往任何时候都清楚见证了划时代的技术创新（机器人、自动化、人工智能、基因工程、后全球化）所带来的进化进程的强劲加速。这些进程带来的挑战不能仅仅归结为信息以及知识的传递。因此，（在管理和战略思维层面上）必须着手修订意义参考框架，在此框架内确定信息和知识交流的坐标。

Task Description:

Today, more than ever, the post-digital society is witnessing a strong acceleration of the evolution processes due to the pandemic and some epochal technological innovations (robotics, automation, artificial intelligence, genetic manipulation, post-globalization). These processes pose challenges that cannot be reduced to information and knowledge transfer alone. It has thus become indispensable (also at the level of management and strategic thinking) to proceed with a revision of the reference framework of meaning within which the coordinates for the exchange of information and knowledge are defined.

指导老师：

Francesco Galli，意大利米兰理工大学设计领导学博士，意大利米兰语言和传播自由大学领导与创新思维教授，研究员，国际化校长代表

Tutor:

Francesco Galli, Ph.D. in Design Leadership from Polytechnic University of Milan, Italy; Professor and Researcher in Leadership and Creative Thinking, IULM University Milan, Italy; Rector's Delegate for Internationalization

反向思考

Think in Reverse

作者：蒋佳玲、梁珊、谷叔玥、杨汀汀、桂煜欣、Janice Florencia Rachmat、石颢沄、赵建钧

By Jialing Jiang, Shan Liang, Shuyue Gu, Tingting Yang, Yuxin Gui, Janice Florencia Rachmat, Haoyun Shi, Jianjun Zhao

今天，由于划时代的技术创新，后数字社会的演化进程正在比以往任何时候都要加速发展。然而，人们在获取这些信息的过程中，并没有进行真正地理解和思考，只是在单一地接收信息。在这个课题的研究中，我们强调面向过去进行反思，从相反的方向探索问题的本质和背后的故事。我们回顾文化和历史，进行哲学批判性思考，寻找未来可行的理论解决方案。

Today, thanks to the pandemic and some epoch-making technological innovations, the evolution of the post-digital society is accelerating more than ever. However, in the process of acquiring this information, people do not really understand and think about it, but only receive information in a single way. In the research of this task, we emphasize reflecting on the past, and exploring the essence and nature of the problem from the opposite direction and the story of back. We look back at culture and history, and engage in philosophical critical thinking to find possible theoretical solutions for the future.

3.2.12　课题：可持续的振动：声音、音乐及视听作品的设计

3.2.12　TASK: SUSTAINABLE VIBRATIONS: DESIGNING SOUND, MUSIC, AND AUDIOVISUAL

课题说明：

　　本工作坊从可持续性的角度在宏观上和跨学科领域探索声音和音乐。它聚焦于声音和音乐设计、创作和制作的技术，以及与可持续性主题相关的声音，聆听与音乐的文化相关联的议题。工作坊涵盖的技术包括声音编辑、分析、合成、声音和音乐创作及其在视听作品中的整合。学员将熟练掌握专业声音和音乐的处理方法、软件和程序。在文化系统中，声音和音乐涉及可持续性的不同方面，例如人类、社会、经济和环境。技术推动了声音、图像、空间和表演的融合，创造了新的合作架构，从而产生了新型的跨国对话。在一个充斥着无数微不足道的事物和视听技术工具的社会中，我们已经被视听印象的洪流和信息洪流所淹没。因此，我们鼓励学生创作出对视听传播带有批判性和反思性的作品，全面阐述声音、音乐和图像之间的关系，超越传统的艺术和设计形式。可持续发展需要解构霸权话语，探索声音、音乐和视听作品形式的多样性，以提升幸福感。

Task Description:

This workshop explores the broad and interdisciplinary field of sound and music from the perspective of sustainability. It examines the technologies of sound and music design, creation, and production along with cultural and critical issues of sound, listening, and music related to the topic of sustainability. The technologies covered in the workshop include sound editing, analysis, synthesis, sound/music composition, and its integration in audiovisual work. Students will become familiar with methods, software, and procedures for working with sound and music at a professional level. Within cultural systems, sound, and music address distinct aspects of sustainability such as human, social, economic, and environmental. Technology propelled convergences of sound, image, space, and performance to create new architectures of collaboration giving rise to new kinds of transnational dialogues. In a society inhabited by myriad trivial objects and gadgets of audiovisual technology, we have become saturated by the torrent of audiovisual impressions and the flood of information that can lead to a state of entropy. Therefore, we will encourage students to create works that develop a critical reflection on audiovisual communication, a comprehensive account of the relation between sound, music, and image beyond the conventional forms of art and design. Sustainability requires deconstructing the hegemonic discourses and exploring the diversity of forms of sound, music, and audiovisual composition to enhance well-being.

指导老师：

　　Paulo C. Chagas，美国加利福尼亚大学河滨分校作曲教授

　　亓梦婕，中国音乐学院教师，中央音乐学院博士后研究员

Tutor:

Paulo C. Chagas, Professor, University of California, Riverside, USA

Mengjie Qi, Lecturer of China Conservatory of Music, Postdoctoral Researcher at Central Conservatory of Music

可持续的振动：声音、音乐及视听作品的设计

Sustainable Vibrations: Designing Sound, Music, and Audiovisual

作者：陈诗燕、粘婉芸、陈伊凌、费小芹

By Shiyan Chen, Wanyun Nian, Yiling Chen, Xiaoqin Fei

共生城市

针对联合国可持续发展目标 11，聚焦城市的交通，从人的角度来观察我们城市生活空间中的不和谐之处。以城市交通噪音为灵感而创作的音乐，和城市交通中最具代表性的元素所提炼出的几何图形相结合，创作出声音可视化的动态图形作品。

脑袋积水

作品主要想表达人的情绪容易因为累积而爆发，提醒人们不要将情绪闷在心里，要表达自己的情绪，不然心容易生病。音乐的开始是水渐渐累积的声音接上嗡嗡声，表现情绪累积到满溢而造成的不适感，中间以雷雨作为情绪爆发的状态，之后采用倒转的手法，表现情绪从满溢回到初始的状态，最后以平静的鸟叫声作为情绪缓和的表现。

声态循环

作品把人的生活圈分成了三个部分，城市、乡间和自然环境，希望大家能意识到，人需要文明的同时也离不开自然，我们的文明与自然共生共存。作品想用声音来表现人与社会环境的联结。如果能落实生态系统的保护，它就像一个强大的泵，将能持续发展，这就是一种可持续的振动。

在场

"在场"用声音和音乐叙事呈现人们的种种感受，比如迷茫、脆弱、自然的恒常与无常……让人们可以通过作品发声，观照自己的内心，将苦痛与压力释放，最终得到精神上的安抚，感受希望的降临。

Symbiosis City

In response to the UN Sustainable Development Goal 11, the task focuses on urban transport and looks at the disharmony. in our urban living spaces from a human perspective place. Music inspired by urban traffic noise is combined with geometric. Figures extracted from the most representative elements of urban traffic to create a sound visualization dynamic graphic work.

Water in the Brain

The work mainly wants to express that people's emotions are easy to explode due to accumulation, reminding people not to keep their emotions in their hearts, but to express their emotions. Otherwise, the heart will easily get sick. At the beginningof the music, the sound of water gradually accumulating is followed by the humming sound, which expresses the discomfort caused by the accumulation of emotions to the overflow. In the middle, the thunderstorm is regarded as the state of theemotional explosion, and then the reverse method is used to express the emotion from overflowing to the initial state. state, and finally with the peaceful bird call as the expression of emotional relaxation.

Sound Cycle

The work divides the human life circle into three parts, the city, the countryside, and the natural environment. I hopeeveryone can realize that while human beings need civilization, they cannot do without nature. Our civilization coexists withnature. The work uses sound to express the connection between people and the social environment. If the protection of theecosystem can be implemented, it will be like a powerful pump, and it will be able to develop sustainably. This is a sustainablevibration.

Presence

"Presence" uses sound and music to narrate the outbreak of people's various feelings about it, such as confusion, fragility, and natural permanence and impermanence... Let people speak through the works, observe their hearts, and put pain and suffering together. Relieve the pressure, and finally get spiritual comfort, and feel the arrival of hope.

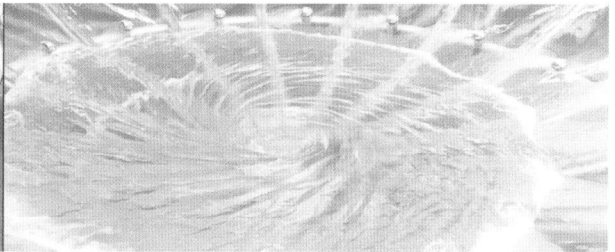

声态循环

voice
cycle

陈伊凌
Yiling Chen

↘ **3.2.13 课题：空间激活的福祉**

3.2.13 TASK: SPACE ACTIVATION FOR WELL-BEING

课题说明：

地点 —— 西澳大利亚州东弗里曼特尔的里奇蒙镇。里奇蒙镇是一个具有混合用途的住宅和商业公寓综合体，位于东弗里曼特尔，该区域有超过三栋建筑和一个广场。里奇蒙镇有 144 个土地板块，119 套住宅公寓，其余为专业或酒店式商业服务场所，包括 Seashells 短期住宿公寓、Sweetwater 屋顶酒吧和烧烤及其他许多场所。这处房产已有五年的历史，三栋建筑之间的广场面积相当大，但是由于全年强风肆虐，这处房产的使用率不高。Strata 公司作为里奇蒙镇的管理者，想要改善这一状况，为社区、业主、房客，以及宾客创造一个充满活力、生机勃勃的空间，可能的措施包括拆除现有建筑物，建造花坛、座椅、游乐设备。学生想要完成这一空间的项目改造必须以一种能促进可持续生活方式的方法进行。

指导老师：

Jake Schapper，澳大利亚科廷大学设计与建筑环境学院城市与区域规划讲师
Anne Farren，澳大利亚科廷大学设计学院教授

Task Description:

SITE—Richmond Quarter, East Fremantle, Western Australia. Richmond Quarter is a mixed-use, residential and commercial apartment complex with over three buildings and a piazza in East Fremantle. Richmond Quarter includes 144 land blocks, 119 residential apartments, and the remaining as professional or hospitality business services including Seashells short stay apartments, Sweetwater Rooftop Bar and Grill and many others. The property is now five years old. The Piazza in between the three buildings is a considerable size, and underused predominantly due to the high winds that occur through the buildings throughout the year. The Strata company, being the managers of Richmond Quarter would like to understand and manage the wind and create a vibrant engaging space for the community, owners, tenants, and guests. This may include the removal of existing structures and the building of Garden beds, seating, and play equipment. The student projects for this space must be created/constructed in a way that fosters sustainable lifestyle choices.

Tutor:

Jake Schapper, Lecturer in Urban and Regional Planning, School of Design and the Built Environment (SDBE), Curtin University, Australia
Anne Farren, Professor, School of Design, Curtin University, Australia

空间激活健康

作者：Jack McMahon、Aram Cho、Yongkang Guan

　　我们的工作始于对缺乏活力和配套基础设施的公共空间进行改造。通过社区咨询，我们了解到社区的需求如下：减缓风吹日晒影响的基础设施、美观和可食用的花园、充足的座位、提供社交空间、负担得起的可回收材料。"亲生物设计"被认为是可以满足多种要求的概念，因此这个概念成为设计的主要策略。

Space Activation for Well-being

By Jack McMahon, Aram Cho, Yongkang Guan

　　Our work begins with the need to regenerate public spaces that lack vitality and supporting infrastructure. Through community consultation, an understanding of community needs was determined as follows: infrastructure to mitigate sun and wind impacts, aesthetic and edible gardens, ample seating, space to provide socializing, and affordable recyclable materials. "Biophilic design" was identified as a concept that could meet many requirements, so this concept was the main strategy of the design.

↘ 3.2.14　课题：可穿戴智能纺织品的未来

3.2.14　TASK: WEARABLE AND SMART TEXTILE FUTURE

课题说明：

　　智能纺织品与服务生态系统的融合程度越来越高，过去十年，以数据这一无形材料扩展进行功能上的延伸。纺织品具有亲和性、柔韧性，对人体和环境具有多种适用性。全球纺织品市场规模预计2021年至2028年的复合年均增长率（Compound Annual Growth Rate, CAGR）为4.4%。智能纺织品可以成为人的第三层皮肤，在环境保护、数据传感器、机器人、可再生能源、导电、温度控制、数字织物——纤维等领域应用广泛。因此，可穿戴的智能纺织品的发展将是人类未来数字生活中解决复杂问题的关键和综合解决方案。

Task Description:

　　Smart textiles are becoming more integrated with service ecosystems and extended material properties are intangible such as data use and change of functionality over the decade. Textiles are human-friendly, flexible, and have various applicability to the human body and environment. The global textile market is expected to expand at a compound annual growth rate(CAGR) of 4.4% from 2021 to 2028. Smart textiles can be the third skin as functional elements for environmental protection, data sensor, robotic, renewable energy, conductive electricity, temperature control, digital fabric-fiber, etc. in broad ways of application. Therefore, the development of wearable and smart textiles will be crucial and integrated solutions under complex problems inhuman's future digital life.

指导老师：

　　朴智瑄，韩国祥明大学公共设计中心特聘教授

Tutor:

Jisun Park, Special Professor, Sangmyung University, Korea

可穿戴和智能纺织未来

Wearable and Smart Textile Future

作者：Muhui Ou, Gyuri Mun, Songhee Han, Yeayeon Choi, Ahhyun Park, Naeun Ha, Yuling Pan

By Muhui Ou, Mun Gyuri, Han songhee, Choi Yeayeon, Park Ahhyun, Ha Naeun, Yuling Pan

随着可穿戴智能纺织品的不断发展，我们的团队为孕妇设计了服装和应用程序。衣服分为两类：上衣和紧身裤。纤维状人造肌肉是这些衣服的主要功能。当它受到电刺激时，它会收缩和放松。在这个过程中，由于纤维的扭曲结构，它能够旋转并产生能量。因此，当纤维的两端固定时，人造肌肉纤维可以通过在身体移动时一起移动来支撑肌肉，以减轻身体和肌肉的压力，这些衣服也可以检查我们的身体状况并将信息发送到电脑或手机。

Since wearable smart textiles have been developing continuously, our team has designed clothes and application for pregnant women. There are two types of clothes: top innerwear and leggings. Fibrous artificial muscle is the main function of these clothes. When it is stimulated by electricity, it contracts and relaxes. In this process, due to the twisted structure of the fibers, it spins and generates energy. So, when the ends of the fiber are fixed, the synthetic muscle fibres can support the muscles by moving together as the body moves to relieve stress on the body and muscles. It can also check our body conditions and send information to a computer or mobile phone.

04

致谢

ACKNOWLEDGEMENTS

4.1 组织

4.1 ORGANIZATION

4.1.1　指导

北京国际设计周组委会
中国纺织服装教育学会
联合国教科文组织国际创意与可持续发展中心

4.1.1　ADVISORS

Beijing International Design Week Organizing Committee
China Textile and Apparel Education Society
International Center for Creativity and Sustainable Development under the Auspices of UNESCO

4.1.2　主办

北京服装学院
北京设计学会

4.1.2　HOSTS

Beijing Institute of Fashion Technology
Beijing Design Society

4.1.3　承办

北京服装学院艺术设计学院

4.1.3　ORGANIZER

School of Art and Design, Beijing Institute of Fashion Technology

4.1.4　协办（按拼音首字母排序）

北京高校学生工作学会
国际计算机音乐协会（ICMA）
国际体验设计大会（IXDC）
美啊设计平台（MEIA）
清华大学艺术与科技创新基地（ATI）
世界幸福城市治理研究中心
西南交通大学国际老龄研究院（NIIA）
新华网融媒体未来研究院（FMCI）
虚拟设计教育论坛（VDEF）
意大利米兰理工大学南京校友会
意大利设计教育协会（IADE）
中国幸福家庭建设研究中心
中欧国际设计文化协会（CEIDA）

4.1.4　CO-HOSTS

Beijing Society for Student Work in Colleges and Universities
International Computer Music Association (ICMA)
International Experience Design Conference（IXDC）
MEIA Design Platform（MEIA）
Tsinghua University Art and Technology Innovation Base（ATI）
Centre of Governance for World Wellbeing Cities
National Interdisciplinary Institute on Aging（NIIA）
Xinhuanet Future Media Convergence Institute（FMCI）
Virtual Design Education Forum（VDEF）
Polytechnic University of Milan Nanjing Alumni Association
Italian Association for Design Education（IADE）
China Family Wellbeing Research Center
China-Europa International Design and Culture Association（CEIDA）

4.2 协办院校（按拼音首字母排序）

4.2 CO-HOSTING SCHOOLS

澳大利亚科廷大学	Curtin University, Australia
奥地利林茨艺术大学	Kunstuniversität Linz, Austria
北京邮电大学	Beijing University of Posts and Telecommunications
北京航空航天大学	Beijing University of Aeronautics and Astronautics
广州大学	Guangzhou University
广州美术学院	Guangzhou Academy of Fine Arts
韩国祥明大学	Sangmyung University, Korea
韩国国民大学	Kookmin University, Korea
丽江文化旅游学院	Lijiang Culture and Tourism College
鲁迅美术学院	Luxun Academy of Fine Arts
马来西亚多媒体大学	Multimedia University, Malaysia
美国旧金山音乐学院	San Francisco Conservatory of Music, USA
美国阿肯色大学	University of Arkansas, USA
美国加利福尼亚大学河滨分校	University of California, Riverside, USA
南京艺术学院	Nanjing University of the Arts
南京信息工程大学	Nanjing University of Information Engineering
葡萄牙马托西纽什艺术与设计学院	ESAD MATOSINHOS, Portugal
清华大学	Tsinghua University
上海立达学院	Shanghai Lida College
上海音乐学院	Shanghai Conservatory of Music
四川大学	Sichuan University
四川音乐学院	Sichuan Conservatory of Music
泰国宋卡王子大学	Prince of Songkhla University, Thailand
中国台湾中原大学	Taiwan Chung Yuan Christian University
天津美术学院	Tianjin Academy of Fine Arts
同济大学	Tongji University
西班牙赫罗纳大学	University of Girona, Spain
西南大学	Southwest University
新加坡南洋理工大学	Nanyang Technological University, Singapore
匈牙利佩奇大学	University of Pécs, Hungary
意大利米兰语言和传播自由大学	IULM University, Italy
意大利米兰理工大学	Politecnico di Milano, Italy
印度尼西亚建国大学	BINUS University, Indonesia
英国新白金汉大学	Buckinghamshire New University, UK
英国格拉斯哥艺术学院	The Glasgow School of Art, UK
英国伦敦米德尔塞克斯大学	Middlesex University London, UK
英国伯明翰城市大学	Birmingham City University, UK
云南艺术学院	Yunnan College of Arts
智利迭戈波塔莱斯大学	Diego Portales University, Chile
中国美术学院	China Academy of Art
中国音乐学院	China Conservatory of Music
中央美术学院	Central Academy of Fine Arts

4.3 学术委员会与组委会

4.3 COMMITTEES

4.3.1　学术委员会主任

贾荣林，北京服装学院校长
詹炳宏，北京服装学院副校长

4.3.2　学术委员（按拼音首字母排序）

Anne Farren，澳大利亚科廷大学教授
Hanny Wijaya，印度尼西亚建国大学副教授
Jisun Park，韩国祥明大学特聘教授
潘荣焕，韩国国民大学教授
常炜，北京服装学院教授
车飞，北京服装学院教授
陈嘉嘉，南京艺术学院教授
丁肇辰，北京服装学院教授
付志勇，清华大学美术学院长聘副教授
宫浩钦，北京航空航天大学教授
郭晓晔，北京服装学院教授
何宇，四川大学教授
胡鸿，北京工业大学教授
黄文宗，台湾中原大学副教授
李若岩，北京服装学院副教授
李政，北京服装学院教授
余春娜，天津美术学院教授
赵璐，鲁迅美术学院教授

4.3.3　总策划

丁肇辰

4.3.4　组委会成员

吕国维、王志国、孙佳琦、宣珂心、卫雨田、司倩、成欣璐、田家荣、潘苗、丁炜航、张嘉琦、卢秋安、范家辉、方智雄、沈鑫、温江亭、邵琪雪、胡一琳、王志国、郝鑫、高昊天、徐晓明、罗梦童、欧阳恩越、聂楠、黄天、刘禹廷、方宇

4.3.1　Academic Chair

Ronglin Jia, President of Beijing Institute of Fashion Technology
Binghong Zhan, Vice President of Beijing Institute of Fashion Technology

4.3.2　Academic Members

Anne Farren, Professor, Curtin University, Australia
Hanny Wijaya, Associate Professor, BINUS University, Indonesia
Jisun Park, Special Professor, Sangmyung University, Korea
Younghwan Pan, Professor, Kookmin University, Korea
Wei Chang, Professor, Beijing Institute of Fashion Technology
Fei Che, Professor, Beijing Institute of Fashion Technology
Jiajia Chen, Professor, Nanjing University of the Arts
Chawchen Ting, Professor, Beijing Institute of Fashion Technology
Zhiyong Fu, Tenure Associate Professor, Tsinghua University
Haoqin Gong, Professor, Beijing University of Aeronautics and Astronautics
Xiaoye Guo, Professor, Beijing Institute of Fashion Technology
Yu He, Professor, Sichuan University
Hong Hu, Professor, Beijing Institute of Technology
Wenzong Huang, Taiwan Chung Yuan Christian University
Ruoyan Li, Associate Professor, Beijing Institute of Fashion Technology
Zheng Li, Professor, Beijing Institute of Fashion Technology
Chunna Yu, Professor, Tianjin Academy of Fine Arts
Lu Zhao, Professor, Luxun Academy of Fine Arts

4.3.3　Executive Director

Chawchen Ting

4.3.4　Committee Members

Guowei Lu, Zhiguo Wang, Jiaqi Sun, Kexin Xuan, Yutian Wei, Qian Si, Xinlu Cheng, Jiarong Tian, Miao Pan, Weihang Ding, Jiaqi Zhang, Qiu'an Lu, Jiahui Fan, Zhixiong Fang, Xin Shen, Jiangting Wen, Qixue Shao, Yilin Hu, Zhiguo Wang, Xin Hao, Haotian Gao, Xiaoming Xu, Mengtong Luo, Enyue Ouyang, Nan Nie, Tian Huang, Yuting Liu, Yu Fang